Neural Network-Based State-of-Charge and State-of-Health Estimation

Neural Network-Based State-of-Charge and State-of-Health Estimation

By

Qi Huang, Shunli Wang, Yujie Wang,
Chao Wang, Carlos Fernandez
and Josep M. Guerrero

Cambridge
Scholars
Publishing

Neural Network-Based State-of-Charge and State-of-Health Estimation

By Qi Huang, Shunli Wang, Yujie Wang, Chao Wang, Carlos Fernandez and Josep M. Guerrero

This book first published 2023

Cambridge Scholars Publishing

Lady Stephenson Library, Newcastle upon Tyne, NE6 2PA, UK

British Library Cataloguing in Publication Data
A catalogue record for this book is available from the British Library

ISBN (10): 1-5275-5217-9
ISBN (13): 978-1-5275-5217-3

CONTENTS

FOREWORD

To deal with environmental deterioration and energy crises, developing clean and sustainable new energy has become the strategic goal of all countries in the world. Lithium-ion batteries are the best solution to provide power and energy storage in the new energy industry at present and are also the main power source of new energy vehicles. State-of-charge (SOC) and state-of-health (SOH) are important indicators to measure whether a battery management system (BMS) is safe and effective. Therefore, this book focuses on the co-estimation strategy of SOC and SOH for power lithium-ion batteries. The book describes the key technologies of lithium-ion batteries in SOC and SOH monitoring and proposes a collaborative optimization estimation strategy based on NN, which provides technical references for the design and application of a power lithium-ion battery management system.

PREFACE

State-of-charge (SOC) and state-of-health (SOH) play a key role in the safety and reliability of batteries. To improve the real-time estimation accuracy of battery state, a novel back propagation neural network-dual extended Kalman filter method for SOC and SOH co-estimation of lithium-ion batteries based on limited memory recursive least square algorithm is proposed. Aiming at the problem of poor anti-interference ability and easy data saturation of recursive least square algorithm, a limited memory recursive least square algorithm that removes the old data and uses new data with only the restricted length is designed to improve the accuracy of on-line parameter identification. Considering the coupling effect between SOC and SOH, the dual extended Kalman filter is used to achieve synergistic estimation to obtain better tracking accuracy. To compensate for the model error of the extended Kalman filter, a back propagation neural network is introduced for correction to further improve SOC and SOH estimation precision. The authors hope that the understanding and experience of predicting the SOC and SOH of a lithium-ion battery can provide some technical reference for the design, matching, and application of power lithium-ion battery management systems, and contribute to the development of new energy technology applications.

CHAPTER 1

INTRODUCTION

With the advantages of small size, high energy density, and low clean pollution, lithium-ion batteries have become an indispensable core energy storage component for the current energy storage industry and the development of new energy vehicles (Xu et al., 2023). State-of-charge (SOC) and state-of-health (SOH) are the core state parameters for the battery management system (BMS) to evaluate and manage the battery state, so it is important to obtain accurate and effective estimates of SOC and SOH in real time to improve the performance of the battery management system and promote the development of electric vehicles.

1.1. Research background and significance

In recent years, as the global climate continues to change, various kinds of extreme weather appear frequently. From torrential rains into disaster and extreme cold, to raging wildfires and high-temperature heat waves, the global "temperature disorder" has destroyed countless lives and homes. To cope with environmental degradation and energy crisis, the search and development of sustainable and clean energy have become a strategic goal for all countries in the world, and carbon neutrality has become a global trend (Cheng et al., 2021; Deng et al., 2022; Zou et al., 2022). China has made a solemn commitment to the international community, "peak carbon emissions and carbon neutrality", striving to reach peak carbon dioxide emissions by 2030 and striving to achieve the strategic goal of carbon neutrality by 2060 (Kong, 2022; Wang et al., 2022c). The White Paper on China's Policies and Actions to Address Climate Change was officially released on October 27, 2021, which is the second time since 2011 that China has released a White Paper on China's response to climate change from the national level (Lu et al., 2023a).

Under the global trend of carbon neutrality, new energy vehicles have seen booming growth (Liu et al., 2022a). Data show that during the five-year period from 2017 to 2021, China's lithium battery shipments continued to climb, maintaining a double-digit increase each year (Zheng et al., 2022b). Among them, the proportion of automotive power battery shipments is rising, accounting for 69% of the national lithium battery market in 2021, far exceeding other application terminals. In the first half of 2022, China's new energy vehicle sales achieved year-on-year growth for six consecutive months, with sales reaching a cumulative total of 2.6 million units, up 120% year-on-year (Tan et al., 2023). At the same time, China's power battery production and installation volume increased significantly by 157.9% and 90.3% respectively. According to the U.S. Department of Energy, 13 new battery gigafactories will come online in the U.S. by 2025 (Ahmed et al., 2023). Another forecast shows that by 2035 Europe may add 35 new super plants for processing lithium resources or producing lithium-ion batteries (Garcia et al., 2023a; Potrc et al., 2021). With the rapid development of new energy vehicles and the increase of intelligent consumption in daily life, lithium battery application scenarios will be more abundant (Zahoor et al., 2023). It is expected that the next 10 to 15 years will be its golden development period. This shows that lithium batteries have broad development and application prospects.

As electric vehicles are becoming more and more accepted, their battery safety issues cannot be ignored (Huang et al., 2021c). Due to the relatively active nature of lithium, spontaneous combustion and explosion accidents related to lithium batteries occur frequently, resulting in consumers' concerns about electric vehicles gradually shifting from "mileage anxiety" to "safety anxiety" (Li et al., 2023a). According to the relevant data survey statistics, from the beginning of 2021 to date the country has occurred more than 50 electric car spontaneous combustion events, the vast majority caused by the battery thermal runaway (Gao et al., 2022b). Power lithium battery safety issues have become a "stumbling block" for the development of the electric vehicle industry. According to research studies, the main concern of Chinese consumers for electric vehicles is safety up to 29%. In addition, from the proportion of complaints about the quality of electric vehicles, battery complaints accounted for 37.1%, including power battery failure, charging failure, battery failure, etc The safety of power batteries has become a new concern for consumers, so it is urgent to further improve the safety of lithium batteries (Shi et al., 2022a; Zhang et al., 2021a).

To solve the battery safety problem and guarantee the safe application of lithium-ion batteries, an efficient, stable, and reliable BMS is crucial (Wang et al., 2022a). As the key core equipment of lithium battery energy storage, BMS integrates function and management to achieve all-round, efficient, and refined management of battery by monitoring battery parameters to estimate battery status to extend battery life, ensure battery safety and improve battery utilization (Adaikkappan and Sathiyamoorthy, 2022; Qays et al., 2022; Wang et al., 2020e). SOC and SOH are key state quantities for BMS evaluation and management, and their accuracy directly affects the safety and effectiveness of the BMS system (Park et al., 2021). SOC characterizes the remaining available capacity of the battery and its accuracy directly affects the mileage range of the electric vehicle. SOH characterizes the safety performance and capacity decay degree of the battery, which is an important parameter for battery fault diagnosis and safety warning, and has an important reference value for battery safety management and life enhancement. Therefore, accurate estimation of SOC and SOH can improve the effective control of lithium-ion batteries by BMS, which in turn can improve the driving range and driving safety of electric vehicles.

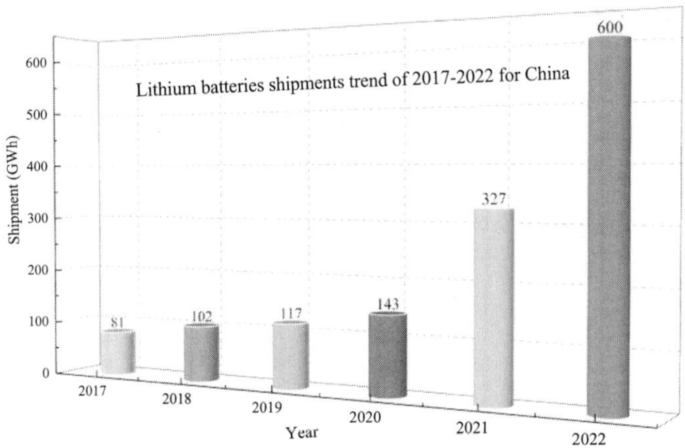

Fig. 1-1 Power lithium-ion batteries shipment trend chart for China

As can be seen visually in Fig. 1-1, from 2017 to 2020, the annual growth rate of lithium battery shipments is maintained at 15 GWh ~ 25 GWh. However, lithium battery shipments are growing rapidly from 2021 onwards, almost doubling every year, with shipments already reaching 600 GWh in 2022. China's lithium battery shipments are expected to exceed 1,000 GWh in 2023, reaching another new level (Hu et al., 2022a).

1.2. Research status

1.2.1. Research status of SOC estimation strategies

SOC is a status parameter that measures the current remaining battery power (Chen et al., 2022; Jiang et al., 2021). An accurate SOC estimate is the key to ensuring battery safety and optimizing battery life level, while providing users with the current battery power status in real time to enhance the user experience (Yang et al., 2022a; Yang et al., 2022b; Zhao et al., 2022). SOC is defined as the ratio of the remaining battery charge under standard discharge rate conditions to the rated capacity under the same conditions, and is defined as shown in Eq. (1-1).

$$SOC = \frac{Q_r}{Q_N} \times 100\% \tag{1-1}$$

In Eq. (1-1), Q_r is the remaining battery power, and Q_N is the rated capacity of the battery under the same conditions, i.e., the current maximum available battery power. The value of SOC ranges from 0 to 100%, when SOC = 100% means the current battery is fully charged, if the battery continues to be charged at this time, the battery may be damaged by overcharging. When SOC = 0, it means the current battery is fully discharged, at this time, if the battery is still discharged, it may damage the battery due to over discharge. Both overcharge and over discharge can cause irreversible damage to the battery, so an accurate SOC value is a prerequisite for preventing overcharge and over-discharge to ensure the safety and health of the battery.

Currently, the estimation methods of SOC mainly include methods based on experimental tests, methods based on model-driven, and methods based on data-driven (Meng et al., 2018; Xu et al., 2021a). The experimental test-based method calculates the lithium battery SOC by measuring physical quantities such as battery voltage, current, and temperature, and then calculating the lithium battery SOC based on known physical relationships, mainly including the discharge method, the open circuit voltage method, and the ampere-hour integration method (Liu et al., 2022b; Wang et al., 2020e; Zhou et al., 2021). The discharge method is generally used as a calibration method for battery SOC estimation or for post battery maintenance work, which is relatively simple, reliable, and has a high accuracy of results (Gao et al., 2022a; Jia et al., 2022). However, the experimental procedure of this method is time-consuming and cannot be used to calculate the battery in operating conditions.

The open circuit voltage method is to obtain the function relationship between OCV and SOC by curve fitting, and then estimate the SOC of lithium-ion battery by OCV-SOC curve (Shan et al., 2022). The method has a simple principle, high operability, and high accuracy of results. However, the measurement of OCV requires a long shelving time of lithium batteries, which is time-consuming and leads to its inability to be estimated online. The ampere-hour integration method calculates SOC based on the integration of current in time. This method focuses only on the external characteristics of the battery system, without considering the internal electrochemical reactions of the battery and the complex relationships between the various parameters (Zhang et al., 2023d; Zhou et al., 2021). Therefore, the calculation is simple and the lithium battery SOC can be estimated online in real time. However, the measurement accuracy of the initial value is required to be high, and the calculation error will gradually become larger with time accumulation because it cannot be corrected in time.

The model-driven approach is based on the working principle and reaction mechanism of lithium batteries to establish the corresponding equivalent model, which is an indirect estimation method with mainly electrochemical models and equivalent circuit models (Shrivastava et al., 2019; Wang et al., 2022e; Wang et al., 2022k). The electrochemical model is based on the principle of electrochemical reaction and describes the internal characteristics of lithium batteries in detail. Its physical meaning is clear and the model accuracy is high, which is suitable for theoretical analysis, but the model structure is too complex and has many parameters, which is extremely computationally intensive (McCarthy et al., 2022; Meng et al., 2018; Peng et al., 2022). The equivalent circuit model consists of circuit elements such as resistors, capacitors, and voltage sources, which are combined in different combinations to form a circuit network to simulate the dynamic reaction process inside a lithium-ion battery (Chen et al., 2021a; Wang et al., 2022d; Zhou et al., 2022). Its model is simple, with few parameters, a small amount of calculation, and good real-time performance, usually combined with parameter identification and state space equations to achieve the prediction and estimation of lithium battery related state

(Adaikkappan and Sathiyamoorthy, 2022; Qi et al., 2022b; Zhang et al., 2021b). The technology is quite mature and very widely used.

The method based on the equivalent circuit model first needs to collect information such as battery voltage, circuit, and temperature through experiments, then establish a suitable equivalent model to construct the state space equation, carry out model parameter identification, and finally use a suitable control theory algorithm to estimate SOC (Wang et al., 2022f; Wang et al., 2022g; Zhang et al., 2022a). Currently, the Kalman Filter (KF) and Particle Filter (PF) are more widely used (Pu et al., 2022; Takyi-Aninakwa et al., 2022a; Wang et al., 2022i). Naseri et al. proposed an enhanced equivalent circuit model based on the Wiener structure to improve the nonlinear capability of capturing lithium-ion batteries (Naseri et al., 2022). The results showed that the accuracy of the EKF algorithm to estimate SOC is improved by 1.5% compared to the conventional second-order equivalent circuit model. Duan et al. used a robust EKF method with correlated entropy loss for SOC estimation to improve the estimation accuracy in non-Gaussian environments, and the results show that the mean square error of the proposed method is reduced by 0.849% (Duan et al., 2020). Qiao et al. proposed a firefly algorithm based on intelligent weight decay to improve the particle wave method for estimating SOC (Qiao et al., 2022). This method avoids the algorithm from falling into local optimum by introducing a linear weighting strategy, and data with different complexity conditions are used to verify the feasibility of the proposed algorithm. The experimental results showed that the root mean square error of the calculation is around 1%.

The data-driven method is based on using data to estimate SOC directly by mining the mapping relationship between the own characteristics of battery measurement parameters such as current, voltage, temperature, and internal resistance and the battery SOC (Cui et al., 2022; Wang et al., 2022h; Zhang et al., 2022c). Its advantage is that it does not need to build a battery model, but it must be backed by high-quality measurement data to obtain accurate estimation results (Liu et al., 2023b; Takyi-Aninakwa et al., 2022b; Wang et al., 2022j). At present, neural networks and deep learning are the most common, and they provide a novel solution for SOC estimation.

Hong et al. proposed a multi-step forward online joint SOC prediction method based on the long short term memory (LSTM) neural network and multiple linear regression algorithm, and the experimental validation showed that the method has good stability, flexibility, and robustness (Hong et al., 2020). To overcome the shortcomings of the black box principle and thus fully exploit the performance of deep learning, Tian et al. proposed to integrate two kinds of domain knowledge into a deep learning-based approach (Tian et al., 2022). First, the voltage and current sequences are decoupled into open-circuit voltage (OCV), ohmic response, and polarization voltage to enhance the input of deep neural networks (DNN). Second, a combined framework is proposed to adaptively fuse SOC estimation results from DNN and short-term ampere-hour integral prediction. The results show that this method can significantly reduce the root mean square error and maximum absolute error of SOC estimation by 30.89% and 64.88%, respectively, with a slight increase in the amount of calculation. To overcome the general deep learning models that ignore the changes between battery cells or focus on short-term estimation, Oyewole et al. proposed a controllable deep transfer learning (CDTL) network for short-term and long-term SOC estimation in the early stages of degradation (Oyewole et al., 2022). The experimental results show that the proposed CDTL outperforms existing deep and transfer learning benchmarks, where the RMSE is improved by 50%~60% and the computation time is reduced by about 39%.

1.2.2. Research status of SOH prediction algorithms

SOH is an important indicator to evaluate the degree of battery aging or decline, reflecting the health of the battery, commonly expressed as a percentage, and the new battery is 100% when it leaves the factory (Long et al., 2023; Yang et al., 2021; Zhang et al., 2022b). The accurate SOH value provides the basis for real-time monitoring of the health information of the lithium-ion battery (Fan et al., 2022; Hu et al., 2022b; Ren et al., 2022a). When SOH is lower than 80%, it reaches the end-of-life standard (Huang et al., 2021a; Xiong et al., 2021). At this time, the battery should be replaced in time to ensure the safe and stable operation of the electric vehicle. There are two main definitions of SOH, namely capacity-based definition and internal resistance-based definition (Huang et al., 2021b; Ren et al., 2022b; Tian et al., 2020).

When capacity decay is used as the definition, SOH is the ratio of the current remaining battery capacity to the rated capacity, which is defined as shown in Eq. (1-2).

$$SOH = \frac{Q_{now}}{Q_{new}} \times 100\% \tag{1-2}$$

In Eq. (1-2), Q_{new} is the rated capacity of the battery when it leaves the factory, and Q_{now} is the capacity at the current moment after the battery has been cyclically charged and discharged, that is, after being used. When the capacity is lower than 80% of the rated capacity at the time of delivery, the battery fails and is no longer used as a power source.

When defined by the increase of internal resistance, the definition of SOH is shown in Eq. (1-3).

$$SOH = \frac{R_{aged} - R_{now}}{R_{aged} - R_{new}} \times 100\% \tag{1-3}$$

In Eq. (1-3), R_{new} is the rated ohmic internal resistance of the battery when it leaves the factory, that is, the ohmic internal resistance when the battery is in a new state, and R_{aged} is the corresponding ohmic internal resistance when the battery reaches the scrapping standard, that is, the ohmic internal resistance when SOH is equal to 80%. R_{now} is the ohmic internal resistance corresponding to the current moment after the battery has been used for some time.

At present, the three main types of SOH estimation methods at home and abroad are experimental test methods, model-driven methods, and data-driven methods (Lipu et al., 2018; Qiu et al., 2022; Shen et al., 2021). The experimental test method obtains the SOH value based on the definition of SOH by directly measuring the characteristic parameters of the characterized power cell such as cell capacity, internal resistance, etc (McCarthy et al., 2022; Tan et al., 2021). The experimental test method is simple but requires strict experimental conditions and complex and accurate experimental equipment, and the measurement cycle is long, which is not suitable for online estimation of SOH, and is difficult for practical application (Zhao et al., 2023). It is generally used as a benchmark value to verify the effectiveness of other related estimation methods.

The model-driven method needs to establish a battery model and obtain parameters such as internal resistance and capacity of the battery through the corresponding model mechanism to realize the estimation of SOH (Park et al., 2020). Commonly used models include equivalent circuit models and electrochemical models. The equivalent circuit model is similar to the SOC estimation method, which usually requires a combination of parameter identification and various filtering algorithms to achieve an online estimation of SOH. Due to better accuracy and convergence, equivalent circuit models are widely used and mature, and the more common ones are recursive least squares, $H\infty$ observer, Kalman filter, and particle filter (Hong et al.; Wang et al., 2020e).

Rossi et al. proposed a GO-EKF algorithm for estimating the SOH value of lithium batteries (Rossi et al., 2022). The method optimally adjusts the covariance matrix of the EKF employing a genetic algorithm to solve the minimization problem of the nonlinear function, to avoid falling into local minima and thus determine the covariance matrix coefficients of the EKF. The experiments show that the method has good accuracy. For the problems of low accuracy and poor convenience of existing SOH estimation methods, Liu et al. proposed an improved cuckoo search particle filter (ICS-PF) algorithm (Liu et al., 2021b). The method proposes an improved cuckoo search particle filter algorithm to estimate capacity attenuation based on the traditional particle filter (PF) and cuckoo search (CS) algorithms by enhancing the search step size and discovery probability. The experimental results show that the maximum error of the method is less than 2%. To solve the problems of insufficient fit of a single health index and difficulty of online measurement of battery internal resistance, Yu et al. proposed a multi-dimensional health indicator (HI) battery state of health (SOH) prediction method (Yu et al., 2023). The method is based on the equivalent circuit model to extract the polarization resistance, polarization capacitance, and initial discharge resistance values through the constant current discharge characteristic curve as health indicators, and the optimal equal voltage drop discharge time selected by the retention strategy genetic algorithm (e-GA) as an indirect health indicator, and finally the PSO-LSTM model is established to realize SOH prediction based on the above health indicators. The experimental results show that the accuracy of this optimization strategy is improved by 0.79% compared to the prediction model with a single health indicator input.

To address the problem of low accuracy of SOH estimation for lithium-ion batteries under complex stress conditions, Yang et al. proposed a method for SOH estimation of lithium-ion batteries using an adaptive dual extended Kalman filter-based fuzzy inference system (ADEKF-FIS) (Yang et al., 2020). The experimental results show that the ADEKF-FIS algorithm effectively improves the prediction accuracy of SOH and has better convergence. Zhu et al. proposed a SOH estimation method for lithium-ion batteries based on the traceless Kalman filter and Improved Unscented Particle Filter (IUPF) (Zhu and Fu, 2021). The experimental results show that the UKF-IUPF algorithm can obtain higher SOH estimation accuracy and faster estimation speed compared with the IUPF method.

Thanks to the development of big data and artificial intelligence, the data-driven method has developed rapidly and is widely used in recent years (Jin et al., 2021a; Shu et al., 2021; Wang et al., 2023b). Its advantage is that it does not need to establish a battery model and clarify the electrochemical mechanism of the battery, but requires high-quality data and high-performance computing power to be used as a support, which is similar to the data-driven method of SOC estimation (Chang et al., 2021; Wang et al., 2021b).

Lin et al. proposed a novel SOH estimation method based on an attentional long short-term memory network (LSTM) with multi-source features (Lin et al., 2023a). The method extracts eight features by analyzing IC, DT, and DTV curves, introduces Wasserstein distance to complement the description of battery aging characteristics, and employs a modified LSTM network structure based on an attention mechanism to estimate the SOH of lithium-ion batteries. The experimental results show that the MAE and RMSE of the proposed LSTM model with local attention mechanism based on multi-source features are less than 0.7%, which has higher accuracy and better robustness. However, LSTM has a large amount of calculation and requires a relatively long training time. The current method of predicting the SOH of lithium-ion batteries by health indicators only considers temporal characteristics and ignores the relationship between different characteristics, especially in the case of limited data in applications. Yao et al. proposed a new graph neural network (GNN)-based SOH prediction framework that considers both temporal features and spatial features of health indicators under limited data (Yao et al., 2023). The experimental results show that the SOH error under this prediction framework is the smallest and the accuracy is higher compared with RNN, LSTM, and CNN-LSTM. To achieve accurate SOH prediction in a unified framework of single-step prediction, multi-step prediction, and long-term prediction, Cai et al. proposed a data-driven approach for lithium-ion battery health state prediction in a unified framework (Cai et al., 2022). The experimental results show that the method achieves a unified prediction framework for lithium-ion batteries with uncertainty characterization, and the effectiveness and superiority are more prominent.

Considering the long training time and large computational effort of some neural networks, some researchers have combined the more traditional neural networks with other algorithms, which not only takes into account the nonlinear advantages of the neural network but also makes up for its defect of a large amount of calculation. A simulated annealing-backpropagation (SA-BP) model is proposed by Xiong et al., which can estimate the long-term state-of-health (SOH) of lithium-ion batteries online by combining them with the battery single-particle (SP) model (Xiong et al., 2023). The experimental results show that the SOH estimation results show significant improvements in various performance evaluation metrics. Xue et al. proposed a joint SOH estimation algorithm based on particle filtering, quantum genetic algorithm (QGA), and generalized regression neural network (GRNN) (Xue et al., 2022). Experimental results based on the NASA dataset and data from real vehicles show that the algorithm has the advantages of high estimation accuracy and low computational cost, which is of great significance to the estimation of SOH of electric vehicle batteries under practical working conditions.

1.2.3. Research status of SOC and SOH co-estimation

Due to the coupling effect between SOC and SOH, it is difficult to guarantee an accurate estimation of the real state of the battery by estimating one state alone (Park et al., 2021; Qi et al., 2022a; Qian and Liu, 2021). Therefore, some researchers have proposed a collaborative estimation strategy for SOC and SOH.

Xu et al. proposed a co-estimation algorithm of SOC and SOH based on an adaptive dual extended Kalman filter (Xu et al., 2021c). The experimental results show that the maximum error of SOC estimation is

2.03%, and the maximum error of ohmic resistance is 15.3%. Ren et al. introduced the forgetting factor into the particle wave algorithm and proposed a dual particle wave algorithm based on the forgetting factor for the co-estimation strategy of SOC and SOH (Ren et al., 2022c). The experimental investigation indicates that the estimation results of SOC and SOH are accurate and the convergence speed is fast. Qiao et al. introduced a bias-compensated recursive least square for parameter identification and proposed a multiple-weighted dual Kalman filtering algorithm (Qiao et al., 2021). The method not only improves the accuracy of parameter identification but also can make full use of the multiple time information to achieve the synergistic estimation of SOC and SOH.

Ma et al. established a fractional second-order RC model that can reflect the main electrochemical behavior of lithium batteries, and realized the model parameter identification through an adaptive genetic algorithm (Ma et al., 2022b). Then the multi-innovative unscented Kalman filter (MIUKF) is used to estimate SOC, and the unscented Kalman filter (UKF) is used to estimate SOH, and finally, a new method for the collaborative estimation of SOC and SOH based on the multi-scale FOMIUKF-UKF is realized. The experimental results show that the proposed method has good adaptability and high accuracy under different test cycles and different aging degrees. To improve the accuracy and convergence speed of SOC estimation, Qiu et al. proposed the backward smoothing square root cubature Kalman filter (BS-SRCKF), and then adopted the multiscale hybrid Kalman filter (MHKF) composed of BS-SRCKF and extended Kalman filter to achieve the joint estimation of SOC and SOH (Qiu et al., 2020). The experimental results show that the algorithm can effectively improve the accuracy of SOC and SOH estimation. Wu et al. employed an improved firefly algorithm to replace the resampling of traditional particle filtering to suppress the particle loss during the execution of the standard particle filtering algorithm (Wu et al., 2022). Then the ohmic resistance is taken as a characteristic parameter of the battery health status, and finally, an IFA-PF algorithm is proposed to realize the joint estimation of SOC and SOH. The experiments show that the joint IFA-PF estimation algorithm has better superiority. Some researchers have proposed a joint SOC and SOH prediction algorithm based on the LSTM neural network, which provides higher accuracy and stability of SOC and SOH estimates compared to predicting one state alone (Jo et al., 2021).

In summary, traditional methods are more applied and mature, such as literature [55]-[57] and literature [85]-[88]. However, most researchers focus only on one aspect of SOC or SOH. Some scholars have designed a fractional order model, adopting an adaptive dual square root cubature Kalman filter (ADSRCKF) to achieve an online estimation of SOC and SOH, and some scholars have proposed a data-driven model framework based on deep learning to estimate SOC and SOH (Gong et al., 2022; Li et al., 2023b). However, this type of method simply realizes the prediction of SOC and SOH without considering the coupling relationship between the two, so there are certain defects in the scheme.

At present, only a few researchers have conducted research on the co-estimation of SOC and SOH. Although some scholars have introduced deep learning and neural network methods for SOC or SOH estimation, such as reference [64]-[66] and literature [95]-[99]. However, these networks have high complexity, poor accuracy and stability, and high application cost. Therefore, considering the complexity and application cost, accuracy, and robustness, combining traditional methods with simple and effective neural networks is a proven solution. In addition, due to the coupling effect between SOC and SOH, the collaborative estimation will definitely become a trend for future research. However, most of the current collaborative estimations are based on traditional methods, such as in the literature [102]-[107].

Since the model defects of the algorithms in the traditional methods can indirectly affect the accuracy of SOC and SOH, a robust back propagation neural network (BPNN)-dual extended Kalman filter (DEKF) method for co-estimation based on a limited memory recursive least square (LMRLS) algorithm, also named an LMRLS-BPNN-DEKF collaborative algorithm estimation model, is proposed in this study. Firstly, a novel LMRLS algorithm is designed to overcome the defects of traditional parameter identification and improve online parameter identification accuracy. Secondly, a dual Kalman filter is constructed to achieve the cooperative estimation of SOC and SOH to obtain better tracking accuracy. Finally, the BP neural network is introduced to compensate for the model error of the Kalman filter to further improve the accuracy and robustness of SOC and SOH.

1.3. Research lines and contents

SOC and SOH are important indicators for evaluating battery status, so accurate and real-time acquisition of both pieces of information is crucial to improving battery life and safety. This book takes the ternary lithium-ion battery as the research object. Firstly, the internal mechanism and charging and discharging characteristics of the lithium-ion battery are studied through experiments to analyze the working principle of the lithium-ion battery. Secondly, a second-order RC equivalent circuit model is established according to the characteristic analysis, taking into account the accuracy and complexity of the model, and a limited memory recursive least squares algorithm is designed for online parameter identification. Then, a dual extended Kalman filter is constructed to realize the cooperative estimation of SOC and SOH. Finally, a BP neural network is introduced to correct the estimation error of DEKF to further improve the accuracy and robustness of the estimation results. Fig. 1-2 shows the research route and main content of this book.

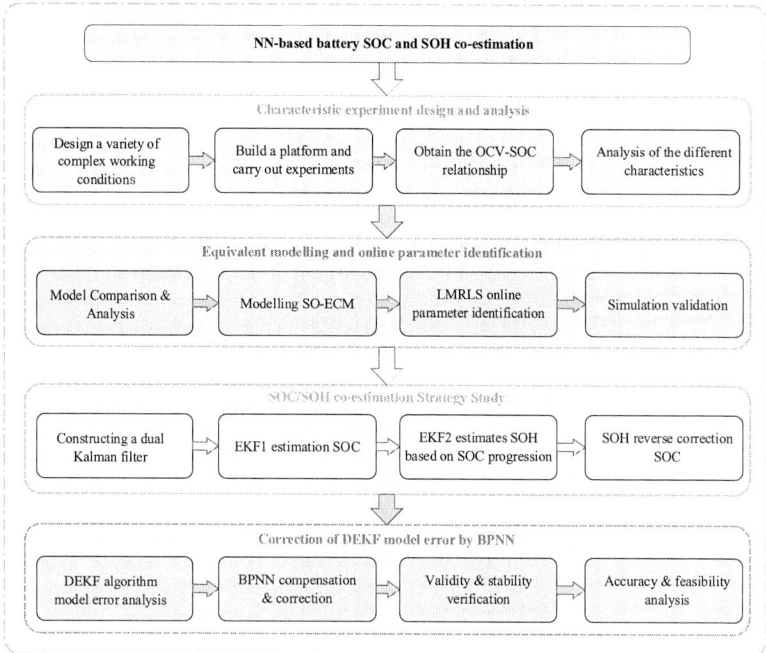

Fig. 1-2 Research route and main content

According to the route framework in Fig. 1-2, the content of each chapter of this book is arranged as follows.

Chapter 1 is the introduction, which analyzes and introduces the research background and significance of this topic. The current status of SOC estimation research, the current status of SOH estimation research, and the current status of SOC/SOH co-estimation research at home and abroad are thoroughly investigated and introduced, respectively. Finally, the shortcomings of the current study are summarized and analyzed, and the main research content and overall research framework of this topic are proposed.

Chapter 2 is the principle and characteristics analysis of power lithium batteries. Firstly, the internal structure and principle of lithium-ion batteries are analyzed in detail. Then, the data of voltage and current at different temperatures are obtained through experiments under various complex working conditions. After extraction and analysis, the effective data is obtained, and then the OCV-SOC function relationship is gained by curve fitting. In turn, the hybrid pulse test characteristics of lithium-ion battery, the variation law of OCV-SOC curve under different temperatures, the variation law of battery capacity under different temperatures, and the variation of terminal voltage under different discharge multipliers are studied and analyzed to provide a basis for battery modeling.

Chapter 3 is a study on the equivalent modeling and parameter identification of power lithium batteries. On the basis of Chapter 2, the second-order RC equivalent circuit model (SO-ECM)) is finally established by analyzing the input and output characteristics of the battery based on the experimental data, obtaining the internal parameters of the lithium-ion battery and its variation law, and taking into account the accuracy and complexity of the model at the same time. To address the problem that the recursive least squares (RLS) algorithm is poorly resistant to interference and prone to data saturation, the limited memory recursive least squares (LMRLS) algorithm is designed that removes the old data while inputting new data and use only the latest data of limited length for parameter estimation, which effectively improves the accuracy of online parameter identification.

Chapter 4 is a study on the cooperative estimation strategy of SOC and SOH based on BP neural network correction. This chapter first analyzes the principle of Kalman filter estimation SOC and its advantages and disadvantages, and discusses the strategy of BPNN optimization EKF estimation SOC, which lays the foundation for the next step of collaborative estimation of SOC and SOH. Then the SOC and SOH synergistic estimation strategy is realized by constructing a dual extended Kalman filter. Where the first EKF implements the estimation of SOC and takes the estimation result as the input of the second EKF for SOH estimation, and then the current moment SOH estimate obtained from the second EKF is used as the input of the next iteration loop to further correct the SOC estimate for the next moment. After the mutual correction feedback of the two forms a closed loop, and the collaborative estimation of the two is finally realized. Finally, the BPNN is introduced to compensate for the model error of DEKF to further improve the accuracy and robustness of the estimation results.

Chapter 5 shows the experimental validation analysis based on the LMRLS-BPNN-DEKF co-estimation model. To verify the effectiveness of the proposed LMRLS-BPNN-DEKF algorithm, detailed experimental validation and analysis of the co-estimation results of SOC and SOH are carried out by using data at different temperatures and complex test conditions, and the estimation results are compared with those of other algorithms. To further verify the robustness and correction capability of the LMRLS-BPNN-DEKF algorithm, experimental validation and analysis are conducted by selecting experimental data at 25°C and setting different initial values of SOC and SOH.

Chapter 6 provides a summary of the book's work and outlook This section summarizes the work and results of this study, analyzes the shortcomings of the current research methodology, and provides an outlook for the subsequent research work in this paper.

1.4. Chapter summary

This chapter first introduces the significance and necessity of this study by introducing global climate change, the shipment of lithium batteries, and the safety of lithium batteries in recent years. Then, the current status of SOC estimation, SOH estimation, and SOC and SOH co-estimation is investigated in detail. Finally, based on the shortcomings of the current SOC and SOH estimation research, the LMRLS-BPNN-DEKF synergistic estimation model is proposed, and the main chapters and contents of this book are briefly introduced.

CHAPTER 2

CHARACTERISTIC ANALYSIS OF POWER LITHIUM-ION BATTERIES

Understanding the internal operating mechanism of power lithium-ion batteries and conducting its characteristic analysis and research has important guiding significance for the subsequent research on battery modeling and synergistic estimation strategies of SOC and SOH. This chapter first briefly introduces the working mechanism of the lithium-ion battery and its internal main structural components, then builds the experimental platform and conducts experiments under complex working conditions, and then performs feature extraction and analysis of relevant data such as current and voltage during charging and discharging, which provides a sufficient scientific basis for the research and discussion of various theories such as battery modeling and parameter identification in Chapter 3 and collaborative prediction system construction in Chapter 4.

2.1 Working mechanism analysis of lithium-ion batteries

Due to many characteristics such as high energy density, excellent temperature adaptability and small clean pollution, and long service life, lithium batteries have been increasingly favored by new energy electric vehicles and various energy storage power stations in recent years (Huang et al., 2022; Zhang et al., 2022e; Zhang et al., 2021c). In this subsection, we will discuss and analyze the internal mechanism and related characteristics of the ternary lithium-ion power battery commonly used in electric vehicles at present. The charging and discharging process of a ternary lithium-ion power battery is carried out through its internal related chemical reactions, so the internal structure and materials should be analyzed and introduced first.

Ternary lithium-ion power battery is mainly composed of four parts, including a positive electrode, a negative electrode, an electrolyte, and an isolation layer (Du et al., 2023a). Among them, the cathode is mainly composed of lithium cobaltate ($LiCoO_2$), lithium iron phosphate ($LiFePO_4$), and ternary substances such as nickel-cobalt-aluminum (Ni-Co-Al) or nickel-cobalt-manganese (Ni-Co-Mn); the main component of the cathode is graphite (Du et al., 2023a; Lu et al., 2023b; Qu et al., 2023). Between the positive and negative electrodes, an electrolyte composed of lithium salts and organic solvents are filled (Yang et al., 2023). The electrolyte mainly assumes the role of a bridge, which is an important medium to ensure that lithium ions can move freely between the positive and negative poles (Hong et al., 2023; Liu et al., 2023a). The isolation layer mainly consists of the isolation film, which is a polymer film with a microporous structure, and is an important inner part of the lithium-ion battery (Song et al., 2021). The diaphragm has two main functions: one is to separate the positive and negative electrodes, like a baffle between the positive and negative electrodes, to avoid short circuits caused by the contact between the two electrodes; the other is to play the role of selective screening – only lithium ions are allowed to pass freely, to form a charge and discharge circuit (Yuan and Liu, 2020). Therefore, the quality of the diaphragm also directly determines the safety, reliability, and effectiveness of the lithium battery. In summary, it can be seen that the internal working mechanism and operation mechanism of lithium-ion battery is shown in Fig. 2-1.

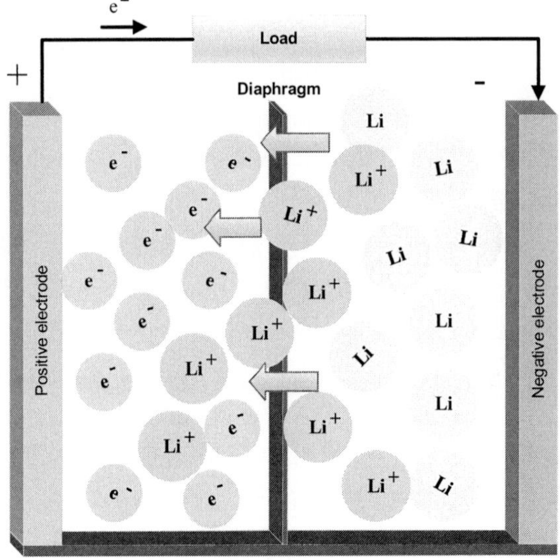

Fig. 2-1 Working principle of lithium-ion batteries

The chemical reaction process inside the lithium-ion battery mainly covers the charging chemistry and the discharging chemistry. As can be seen from Fig. 2-1, when the lithium ions are moving freely back and forth between the positive and negative poles of the battery, the lithium battery generates current through the load connecting the two poles to complete charging and discharging. Specifically, when a lithium battery undergoes a charging chemical reaction, the lithium ions detach from the side of the positive material (usually transition metal oxide), and then the free-floating lithium ions pass through the medium of the electrolyte by free motion and cross the exclusive channel of the isolation film to the negative material (usually graphite) and enter its micropores. At the same time, outside the battery, electrons are withdrawn from the positive material by an external power source, and then enter the negative material via an external circuit, causing the positive material to be oxidized (oxidation reaction occurs). At this point, the electrical energy is gradually converted to chemical energy.

In this process, the concentration of lithium ions in the electrolyte inside the battery gradually decreases, while the concentration of lithium ions in the negative electrode material gradually increases until the battery is in a full state. As the concentration of lithium ions gathered in the negative electrode becomes larger and larger, the corresponding charge into the battery will be higher and higher.

When the lithium battery is discharged, the lithium ions are gradually released from the carbon micropores of the negative electrode material, and then pass through the barrier film utilizing the free movement of the particles, and finally reach the positive electrode material and enter its micropores. At the same time, outside the battery, electrons flow from the negative material to the positive material through the external load circuit, so that the positive material is reduced. At this time, the chemical energy inside the battery is gradually converted into electrical energy. During this process, the concentration of lithium ions in the electrolyte inside the battery gradually increases, while the concentration of lithium ions in the negative electrode material gradually decreases until the battery charge is zero. In summary, lithium-ion batteries work in the process of internal charging and discharging chemical reactions are essentially the same, the lithium-ion through the electrolyte and the diaphragm between the two poles to do round-trip free movement occurs as a result of redox. The above elaborates on the working mechanism inside the lithium-ion battery and lays the foundation for the subsequent analysis of the relevant characteristics of the lithium-ion battery under operating conditions.

2.2. Platform construction and experimental design

This section mainly introduces the experimental platform of the power lithium-ion battery and designs the experiment of complex operating conditions at different temperatures, which lays the foundation for the parameter identification simulation verification in Chapter 3 and the

LMRLS-BPNN-DEKF collaborative prediction model verification in Chapter 5.

2.2.1. Experimental platform

The experimental platform consists of a charge/discharge tester (BTS750-200-100-4), a thermostat box (BTT-331C), and a computer, which connects the battery and the equipment through wires and network cables, as shown in Fig. 2-2.

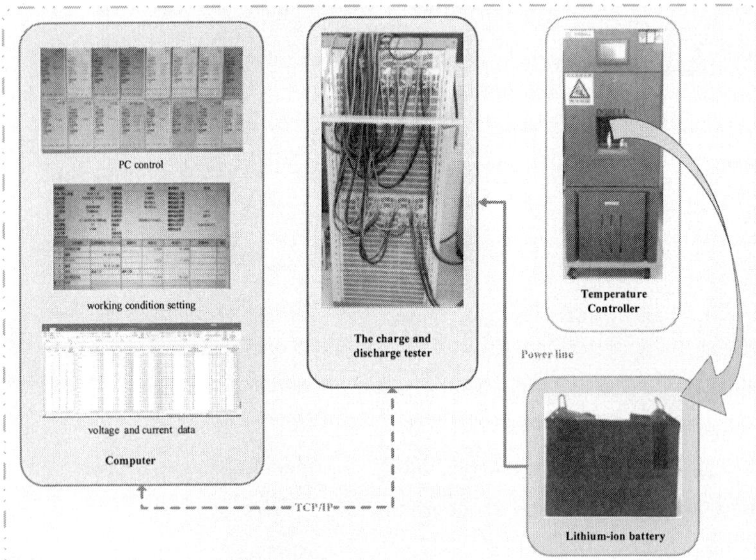

Fig. 2-2 Experimental test platform

The experimental platform is used to set the working condition information and record the voltage, current, capacity, and other data at different moments to get the working condition experiments under different conditions. The complex working condition experiments in this section mainly include the hybrid pulse power characterization (HPPC) test, custom dynamic stress test (DST), and Beijing bus dynamic stress test (BBDST).

2.2.2. HPPC experiments at different temperatures

A HPPC experiment is a standardized test experiment for power lithium-ion battery performance evaluation, which can reflect the characteristics of power battery pulse charge and discharge performance. Through data processing of current and voltage data of HPPC test data, the ohmic resistance, polarization impedance, and the relationship between OCV and SOC of the cell can be obtained for further research related to lithium batteries. The experimental steps of the HPPC test are as follows:

(1) A battery with a current unknown remaining charge is first charged with constant current and voltage (4.2V/1C) until it reaches a fully charged state.

(2) Leave the battery for 40 minutes, and measure and record the voltage value at both ends of the battery.

(3) Perform current pulse experiments on the battery. The battery is discharged at a constant current of 1C for 10 seconds, then left for 40 seconds, and then charged at a constant current of 1C for 10 seconds to restore the battery's charge before discharge.

(4) A constant current discharge of 1C for 6 minutes to reduce the battery charge by 10%, then leave it for 40 minutes and record the terminal voltage as the open circuit voltage at this time.

(5) Repeat steps (3) and (4) 10 times with an experimental interval of 40 minutes between adjacent pulses (the battery is discharged at 1C current for 6 minutes and the SOC value decreases by 0.1) until the battery charge is 0.

Through the above steps, the HPPC operating voltage and current at 15°C, 25°C, and 35°C can be obtained as shown in Fig. 2-3.

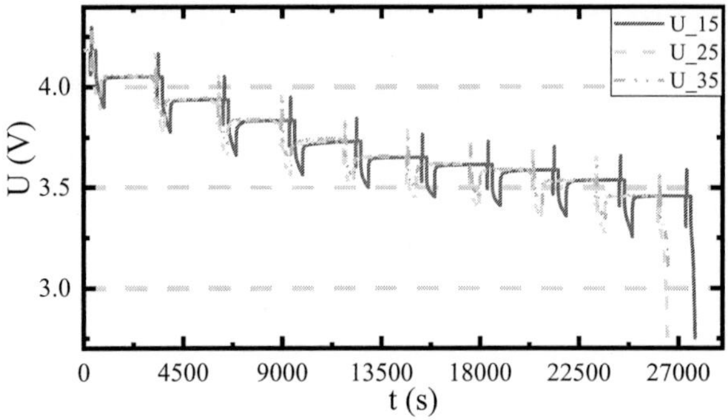

(a) Voltage curve

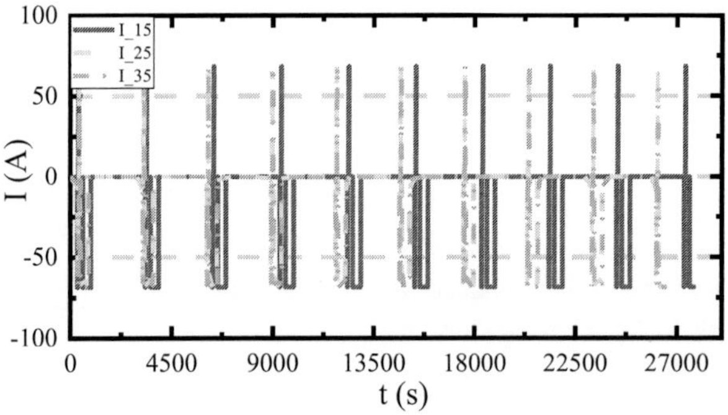

(b) Current curve

Fig. 2-3 Experimental data of HPPC working conditions at different temperatures

As can be seen from Fig. 2-3 (a), as the temperature increases (decreases), the ohmic internal resistance of the battery decreases (increases), resulting in a gradual increase (decrease) in the open circuit voltage of the battery after sufficient resting.

2.2.3. DST experiments at different temperatures

The real-time current of automotive power lithium battery in practical applications is complex and variable, often accompanied by sudden current switching and stopping under different operating conditions, which puts forward strict requirements on the dynamic performance of the battery and brings difficulties to the estimation of SOC and SOH of lithium battery under complex operating conditions. To further improve the closeness of the conducted experiments to the real situation, a custom DST experiment is designed and the specific experimental steps are described as follows:

(1) Charge the battery with constant current and constant voltage, first charging at 1C rate constant current to the upper cut-off voltage 4.2V, followed by constant voltage charging and the charge cut-off current 0.05C.

(2) Setting the battery aside for 30 minutes after charging is complete to stabilize the battery voltage.

(3) The battery is discharged at a rate of 0.5C for 4 minutes at a constant current, and when that current is terminated, the battery is left to stand for 30 seconds.

(4) Charge the battery at a rate of 0.5C for 2 minutes at a constant current, and leave the battery for 30 seconds when it stops charging.

(5) Charge the battery at a rate of 0.5C for 2 minutes at a constant current, and leave the battery for 30 seconds when it stops charging.

(6) Cycle through steps (4) to (5) above until the battery is discharged to 0.

Through the above steps, the DST operating voltage and current at 15°C, 25°C, and 35°C can be obtained as shown in Fig. 2-4.

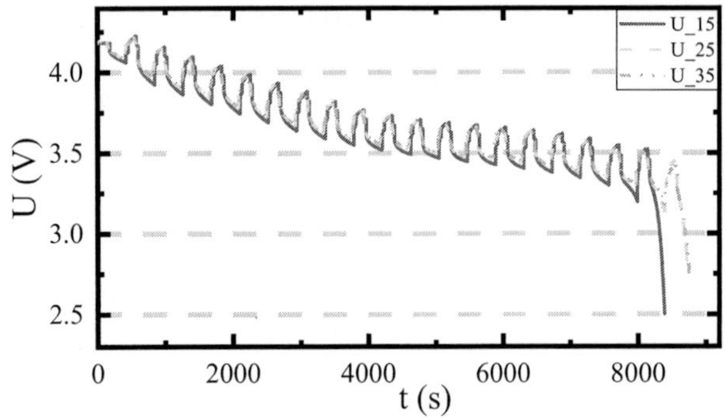

(a) Voltage curve

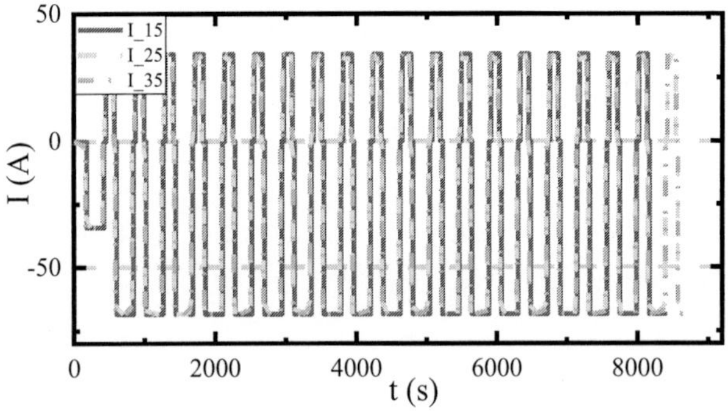

(b) Current curve

Fig. 2-4 Experimental data of DST working conditions at different temperatures

2.2.4. BBDST experiments at different temperatures

To better simulate the dynamics, variability, and irregularity of working conditions in the actual operation of new energy vehicles, the BBDST experiments are conducted on ternary lithium-ion batteries. BBDST is the data collected from the actual operation of the Beijing bus, including not only basic operations such as starting, braking, and stopping, but also accelerating, coasting, and rapid acceleration, which is a more comprehensive and complex test of the power lithium battery. This is also closer to the actual new energy vehicle operation, which lays the foundation for better validation of the model and algorithm. Among them, one small cycle of BBDST working condition is shown in Table 2-1, and one BBDST working condition is composed of several small cycles.

Table 2-1 Description and parameter settings of BBDST working condition

***Ph* (kW)**	***Pc* (w)**	**Single step (s)**	**Total (s)**	**Working condition**
37.5	69	21	21	start
72.5	135	12	33	accelerate
4.5	9	16	49	slide
-15	-27	6	55	brake
37.5	69	21	76	accelerate
4.5	9	16	92	slide
-15	-27	6	98	brake
72.5	135	9	107	accelerate
92.5	174	6	113	rapidly accelerate
37.5	69	21	134	accelerate
4.5	9	16	150	slide
-15	-27	6	156	brake
72.5	135	9	165	accelerate
92.5	174	6	171	rapidly accelerate
37.5	69	21	192	accelerate
4.5	9	16	208	slide
-35	-66	9	217	brake
-15	-27	6	229	brake
4.5	9	71	300	stop

In Table 2-1, *Ph* (kW) is the battery output power under real bus starting, acceleration, coasting, and other working conditions. Since the experiment is conducted on a 70Ah battery, the data in *Pc* (W) is obtained by reducing *Ph* (kW), and *Pc* (W) is the data used for the experiment conducted on this ternary lithium battery. With the above parameters and step settings, the BBDST operating voltage and current at 15°C, 25°C, and 35°C can be obtained, as shown in Fig. 2-5.

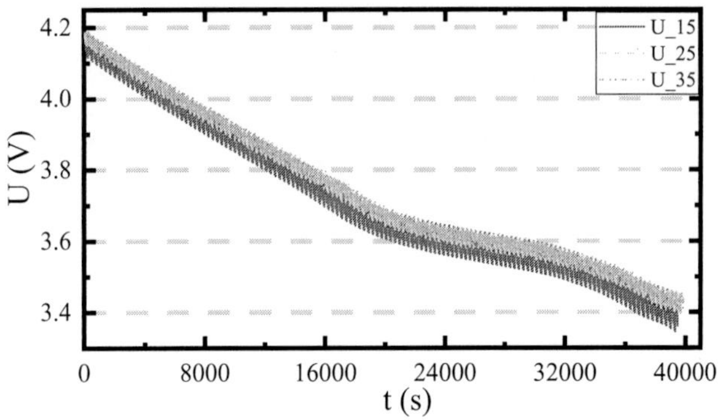

(a) Voltage curve

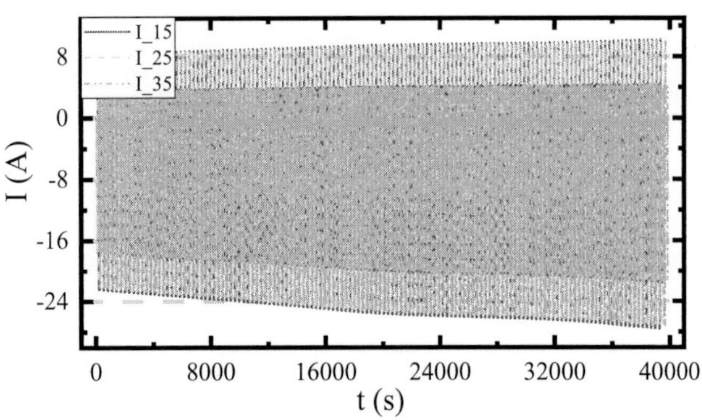

(b) Current curve

Fig. 2-5 Experimental data of BBDST working conditions at different temperatures

2.3. Working characteristic analysis of lithium-ion batteries

Obtaining the operating characteristics and key parameters variation law of the power lithium battery is the basis for the subsequent battery equivalence modeling and state co-estimation. In this section, the hybrid pulse test characteristics of lithium-ion batteries, OCV characteristics at different temperatures, battery capacity characteristics at different temperatures, and discharge characteristics at different multipliers are studied and analyzed.

2.3.1. HPPC test characteristics analysis

The hybrid pulse power characterization test, namely the HPPC test, is a basic experiment of lithium-ion batteries. HPPC analyzes battery performance by applying a series of pulsed currents to the power cell and then measuring the voltage response of the cell. Its main purpose is to determine parameters such as battery capacity and internal resistance and is often used as an experimental validation condition for models and algorithms. The experimental test results in HPPC are shown in Fig. 2-6, where Fig. 2-6 (a) shows the voltage response curve corresponding to the whole HPPC cyclic test process, and Fig. 2-6 (b) shows the voltage response curve of a single HPPC test (at SOC=0.7).

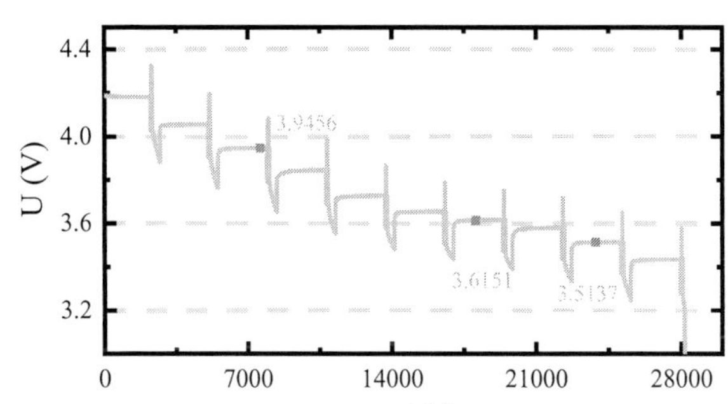

(a) HPPC test the corresponding voltage curve

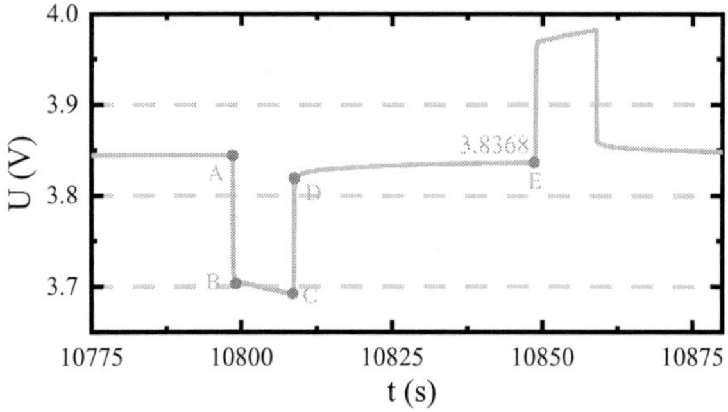

(b) Voltage response curve at SOC = 0.7

Fig. 2-6 HPPC test characteristic curve analysis

As can be seen from Fig. 2-6 (a), overall, the lithium battery shows a regular decreasing trend in voltage during the HPPC test until the cut-off voltage stops changing. Specifically, after the end of each constant current discharge, the lithium battery is put aside for 40 minutes, and its voltage showed a significant change. It can be analyzed from this phenomenon that the voltage of the battery will gradually become stable after being placed for a long time, which indicates that the chemical reaction inside it has become stable, and the voltage value measured in the current state is the open circuit voltage (OCV), and SOC at this time dropped by 0.1. This method is the basic way to measure open circuit voltage. Therefore, the open circuit voltage values that the corresponding SOC (0.1, 0.2, ...,1) can be obtained, and then the dynamic change relationship between the OCV and SOC can be gained by curve fitting, that is, the OCV-SOC curve.

Further analysis of the single voltage response curve from Fig. 2-6 (b) shows that the voltage starts to discharge in the AB section, at which time the voltage drops instantaneously from point A to point B. After that, the voltage shows a slow downward trend (that is, section BC). And then the voltage rises rapidly in a relatively short period of time (that is, the CD section). Subsequently, the battery is maintained in a stable state for a relatively long period of time (that is, the DE segment), which is the resting time period of the lithium battery. The battery voltage is relatively stable during this period of time, and the final stable voltage remains constant as the open circuit voltage. The OCV-SOC fitting curves can be obtained by multiple pairs of OCV and the current corresponding SOC values, laying the foundation for further in-depth study of various state estimates of lithium-ion batteries. The characteristics of OCV-SOC will be discussed in detail in the next subsection.

2.3.2. OCV characteristic analysis at different temperatures

The open circuit voltage (OCV) of a lithium battery generally indicates the terminal voltage value of the battery after sufficient resting (Bao et al., 2022a; Shi et al., 2022b; Yu et al., 2022b). Under the conditions of cell current passage, the change of potential or departure from the equilibrium potential occurs is called electrode polarization (Lin et al., 2023b; Wang et al., 2021c). Electrode polarization can be divided into differential polarization and electrochemical polarization (Li et al., 2021b; Somasundaran et al., 2023). Due to a combination of ohmic and polarization effects, the voltage produced by the battery during discharge tends to change very dramatically in a very short period of time. This can be seen in Fig. 2-6 from the previous section, so the battery is often set aside for a very sufficient amount of time to obtain an accurate OCV value.

The relative research shows that there is a specific law between OCV and SOC (Natella et al., 2023), so the OCV-SOC function relationship can be obtained by the curve fitting method, and then an in-depth study of parameter identification in Chapter 3 and state prediction in Chapter 4 can be carried out. Moreover, the polynomial coefficient of OCV about SOC is also a key debugging parameter for the experimental verification of the simulation in Chapter 5, and its accuracy will directly affect the accuracy of the state estimation results. The actual working environment of lithium-ion batteries is complex and variable. To simulate the real ambient temperature change, experiments are conducted to test the lithium-ion battery at -20°C, 15°C, 25°C, 35°C and 40°C to obtain the OCV-SOC characteristic curves of lithium-ion battery at different temperatures, as shown in Fig. 2-7.

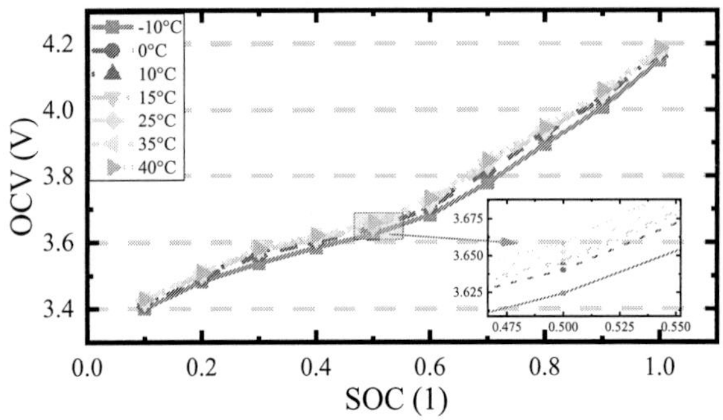

Fig. 2-7 OCV-SOC characteristic curves at different temperatures

It can be seen from Fig. 2-7, horizontally, the OCV decreases with the decrease of SOC at different temperatures, and the decrease is faster when SOC is larger. Longitudinally, when the temperature increases, the OCV is higher at the same SOC. This is due to the fact that the increase in temperature accelerates the chemical reactions inside the battery and plays a catalytic role, which leads to the release of more power from the battery. Therefore, the above analysis shows that temperature has a certain effect on the OCV of lithium batteries and there is a certain pattern.

2.3.3. Battery capacity characteristics at different temperatures

Battery capacity refers to the total power that a lithium battery can discharge when it is fully charged (Park et al., 2018). It is generally represented by the symbol Q, and the unit is mAh (milliampere hours) or Ah (ampere hours). Battery capacity is an important indicator parameter to measure the quality of a battery, and is closely related to the length of time the battery is used (Xu et al., 2021b). Generally speaking, as the battery is used, its capacity will show a slow trend of decline (Liu et al., 2022c). Battery capacity is divided into rated capacity (calibrated capacity) and actual capacity (available capacity). The rated capacity is the minimum amount of electricity that the battery can discharge under certain discharge conditions (temperature, discharge rate, etc.) in accordance with the relevant regulations of the relevant state departments. The actual capacity is the maximum capacity discharged from a fully charged state to the cut-off voltage of the battery under specific conditions. The laboratory generally uses the constant current or constant voltage discharge method to discharge the lithium battery. The actual capacity of the battery (storable power value) is obtained by measuring the amount of electric energy released by the battery during the entire discharge process or calculating the product of the discharge current and time. To obtain the real capacity of the lithium-ion battery, it is necessary to conduct a capacity calibration experiment, the specific steps are as follows:

(1) First charge the battery with a constant current at a rate of 1C, the charging cut-off voltage is 4.2V, and then switch to constant voltage charging, the charging cut-off current is 0.02C, to ensure that the battery is fully charged.

(2) Let the battery stand for 30 minutes.

(3) Discharge the lithium battery with a constant current at a rate of 1C, and the discharge cut-off voltage is 2.75V.

(4) Let the battery stand for 30 minutes.

Repeat steps (1)-(4) three times to end the experiment. Take the average of the three discharge capacities to obtain the actual capacity of the battery in the current state. Fig. 2-8 is the capacity change curve of the lithium-ion battery in the capacity calibration experiment.

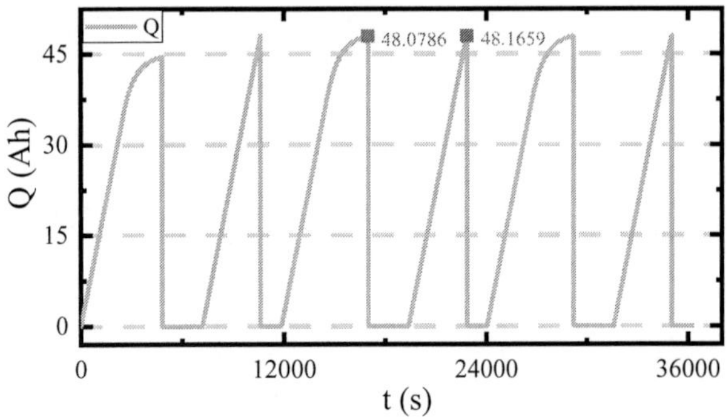

Fig. 2-8 Capacity change curve in calibration experiments

Since temperature tends to have a more pronounced effect on the chemical reaction, the higher the temperature is, the more adequate the chemical reaction is, in theory. And the internal reaction mechanism of lithium-ion battery is closely related to the chemical reaction, so the battery capacity will also vary depending on the external temperature To explore the influence of the relationship between temperature and battery capacity, it is necessary to conduct capacity calibration experiments on lithium batteries under different temperature conditions to obtain the actual capacity of lithium batteries at different temperatures corresponding to the current state, so as to obtain the law of change of lithium battery capacity and temperature, as shown in Fig. 2-9.

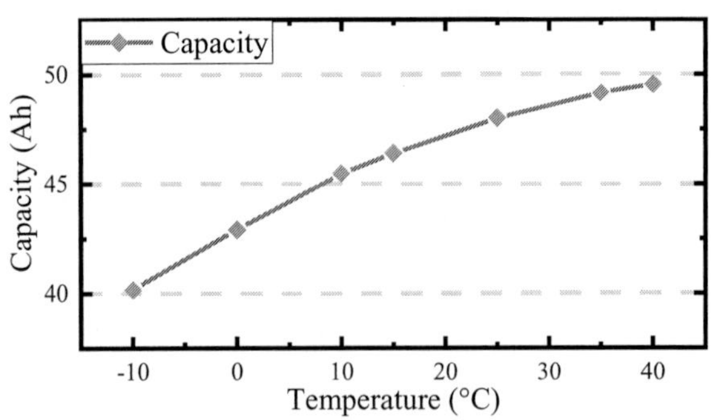

Fig. 2-9 Capacity characteristics of lithium batteries at different temperatures

From Fig. 2-9, it can be seen that the capacity change of the lithium-ion battery will keep rising (falling) as the temperature rises (falls), and when the temperature reaches a certain limit, the battery capacity tends to stabilize. This is due to when the temperature is higher, the activity of the internal materials of the battery is enhanced, which leads to faster chemical reactions, thereby increasing the battery capacity. A reasonable temperature range will enable the lithium battery to give full play to its performance. Once the maximum temperature to which the lithium battery is adapted is exceeded, it will cause irreversible damage to its structure, such as damage to the internal diaphragm, rupture of the shell, and other hardware problems, thus increasing the risk of battery leakage and fire.

2.3.4. Discharge rate characteristics analysis of lithium-ion batteries

The discharge rate of a lithium battery is used to indicate the size of the discharge current, which refers to the size of the ratio between the working current of the battery and the rated capacity of the battery during the discharge process (Liu et al., 2022d; Liu et al., 2021c). It is represented by C, and the rated capacity of the lithium battery can be estimated by different discharge currents (Zheng et al., 2022a). For example, 1C discharge means that the discharge current of the battery is 1 times its rated capacity, that is, the current at this time is equal to the battery capacity in value. Specifically, if a battery has a capacity of 3000mAh, its 1C discharge current is 3000mAh. If the discharge current of the battery is 1500mAh, its discharge rate is 0.5C The physical and chemical reaction process inside the lithium battery is affected by various factors such as electrode materials, external temperature, and battery discharge rate. To qualitatively study the influence of the discharge rate on the battery, the battery is discharged at different rates (I=1.5C, 1.2C, 1.0C, 0.8C, 0.5C) at room temperature (25°C). The output voltage is sampled at a sampling frequency of 0.1s, and the terminal voltage variation curves are obtained under different discharge rates, as shown in Fig. 2-10.

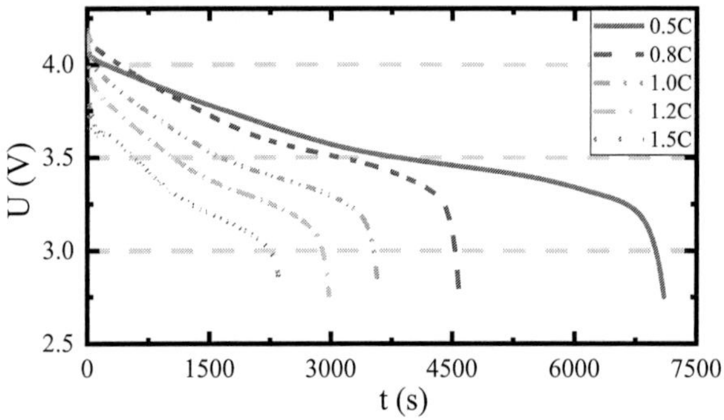

Fig. 2-10 Terminal voltage curve at different discharge rates

It can be seen from Fig. 2-10 that during the entire constant current discharge process, the change of the battery terminal voltage presents a nonlinear characteristic. The first stage is the initial stage of discharge, and the voltage drops rapidly with the sudden change of current. This sudden change is due to the transient ohmic polarization effect occurring inside the battery, and then the instantaneous decrease in terminal voltage caused by the partial voltage of the ohmic internal resistance. The second stage is the "smooth period", when the terminal voltage curve tends to be relatively stable. At this time, the electrochemical reaction inside the battery is relatively stable, and the voltage difference between the positive and negative electrodes decreases slowly. The third stage is the end of discharge, when the battery discharges to around 3.2V, the slope of the terminal voltage curve changes suddenly again, and the slope is almost 0, and lasts until the discharge cut-off voltage. To sum up, it can be seen that the change of the terminal voltage during the constant current discharge process is divided into three stages: "steep-smooth-steep". With the continuous increase of the discharge rate, the battery discharge time is gradually shortened. The larger the rate, the faster the curve declines, and the shorter the "smooth period". It should be noted that although increasing the discharge rate can shorten the discharge time, too large a discharge rate can easily cause excessive battery temperature, which in turn can cause irreversible damage to the battery.

2.4. Chapter summary

This chapter first analyzes the internal structure and working mechanism of lithium-ion batteries in more detail, so that we have a deeper understanding of its internal electrochemical principle, which will help us to dig deeper into the analysis of the factors affecting the working state of lithium-ion batteries. Then, the hybrid pulse test characteristics are explored, the OCV-SOC characteristics at different temperatures, the variation of the battery capacity at different temperatures, and the variation of the terminal voltage at different discharge rates. HPPC test is conducted for studying OCV-SOC characteristics, which are a necessary condition for further simulation experiment verification. The battery capacity is the premise to study SOH, and the charge-discharge rate is the necessary basis for carrying out scientific experiments and collecting effective data. Therefore, the above research and analysis will provide a theoretical basis and design ideas for subsequent battery modeling, parameter identification, and state estimation.

CHAPTER 3

EQUIVALENT MODELING AND PARAMETER IDENTIFICATION OF BATTERIES

Establishing an accurate equivalent circuit model is a prerequisite and foundation for achieving accurate battery state estimation. In this chapter, by comparing the advantages and disadvantages of different models, we finally adopt the second-order RC equivalent circuit model that can characterize the nonlinear relationship of the lithium battery, which considers both the computational power and accuracy of the model. To solve the problem that the recursive least squares (RLS) algorithm has the poor anti-interference ability and is prone to data saturation, a recursive least squares algorithm with limited memory is designed to effectively improve the accuracy of online parameter identification. Finally, the parameters of the designed algorithm model are identified and verified under different temperatures and complex working conditions.

3.1. Second-order model and state space equation

The equivalent circuit model employs lumped elements such as resistors, capacitors, and voltage sources to describe and capture the dynamic behavior and polarization characteristics inside the battery (Chen et al., 2023; Tran et al., 2021; Zhang et al., 2022a). Compared with the electrochemical model, the equivalent circuit model not only characterizes the electrical properties of the battery, which are easy to understand, but also abundant circuit elements and their combinations, with better convenience and flexibility.

3.1.1. Comparative analysis of different equivalent circuit models

The general equivalent circuit models include the Thevenin model, second-order RC model, PNGV model, and GNL model (Ji et al., 2021; Meng et al., 2018; Wang et al., 2020e), as shown in Fig. 3-1.

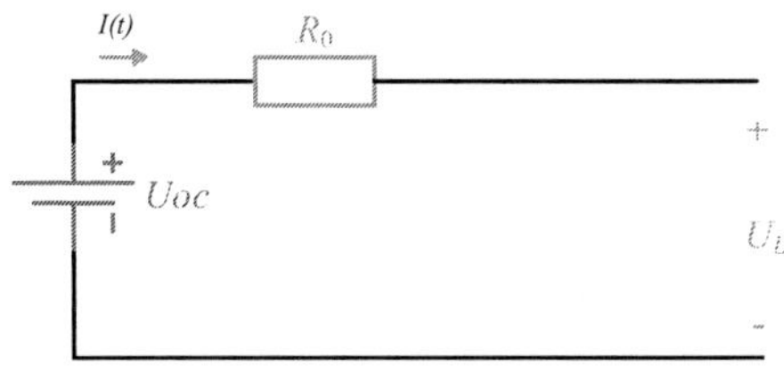

(a) Rint model

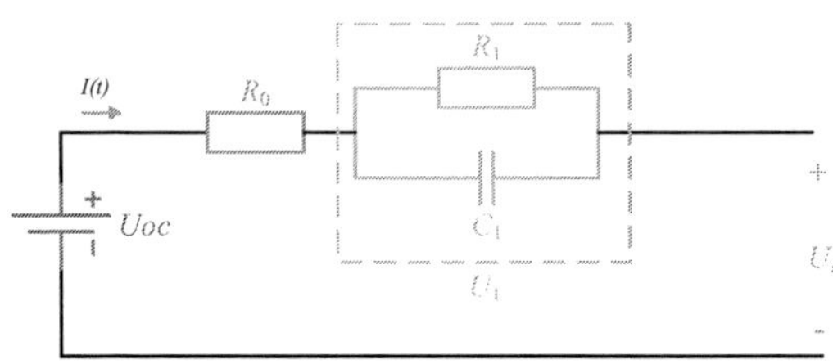

(b) Thevenin model

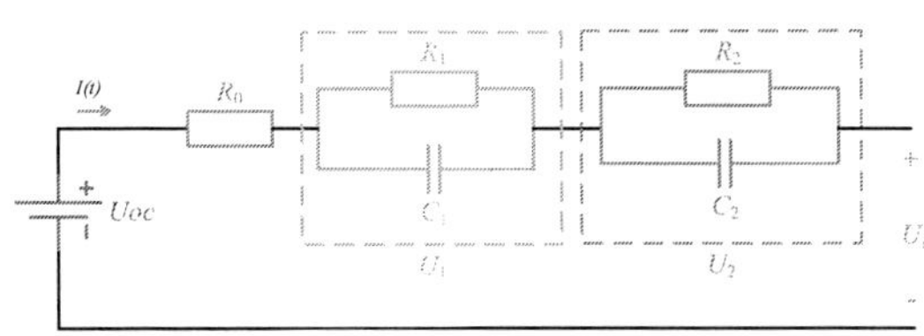

(c) Second-order RC model

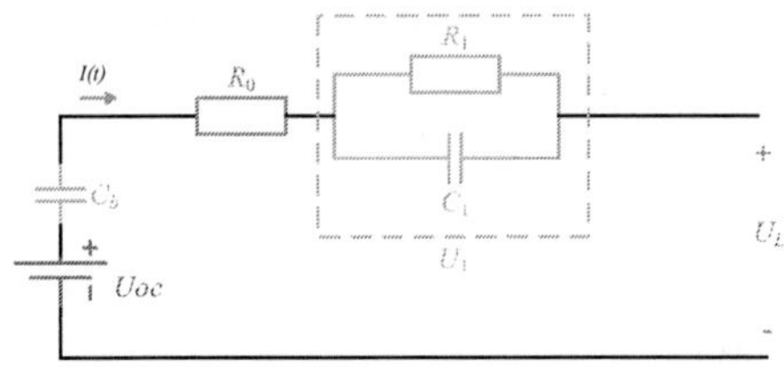

(d) PNGV model

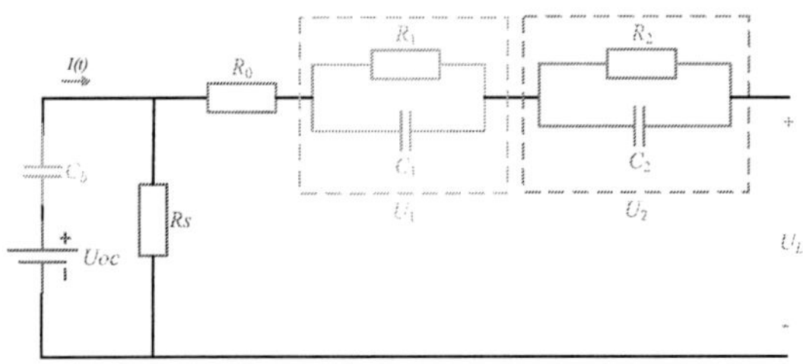

(e) GNL model

Fig. 3-1 Equivalent circuit model

The Rint model, also known as the internal resistance model, is shown in Fig. 3-1 (a), which only considers the ohmic effect of the battery and is the simplest but also the least accurate. The Thevenin model and the PNGV model are first-order equivalent circuit models. The Thevenin model is also known as the first-order RC model, as shown in Fig. 3-1 (b). Based on the Rint model, a parallel RC circuit is added to represent the polarization effect of the battery, which can better characterize the dynamic characteristics of the battery. On the basis of the Thevenin model, a series capacitor is added to the PNGV model to characterize the OCV of lithium-ion batteries with the change of current time integral, as shown in Fig. 3-1 (d). It can reflect both the capacity of the battery and the DC response characteristics of the battery, which is more accurate than the Thevenin model. The second-order RC model and GNL model are the second-order equivalent circuit models. In the second-order RC model, a parallel RC circuit is added based on the

Thevenin model, as shown in Fig. 3-1 (c), so that the electrochemical polarization reaction and concentration polarization reaction of the battery can be characterized by two RC circuits respectively, which further improves the accuracy of the model. Based on the second-order model, the GNL model introduces an RC series network to consider the effect of battery self-discharge, as shown in Fig. 3-1 (e), which can simultaneously simulate the ohmic polarization, electrochemical polarization, concentration polarization, and battery self-discharge phenomenon. Since it is optimized and improved based on the second-order RC model, the accuracy of the model is higher but the complexity also increases.

In summary, considering the difficulty of identification and the amount of calculation, the second-order RC equivalent circuit model which can accurately characterize the polarization and dynamic behavior of lithium batteries is adopted in this study.

3.1.2. Building second-order model and state space equation

The second-order RC model takes into account the accuracy and complexity of the model, and can better reflect the relationship between the electrochemical reaction process of the battery and the concentration of electrode solids, electrolyte concentration, and open-circuit voltage (Bobobee et al., 2023; Lai et al., 2018). Fig. 3-2 shows the circuit model.

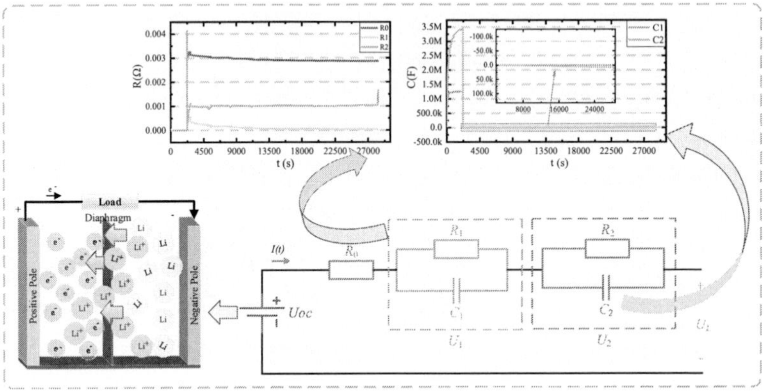

Fig. 3-2 Second-order RC equivalent circuit model

In Fig. 3-2, U_{OC} represents the open circuit voltage of the battery, U_L represents the measuring terminal voltage (the closed-circuit voltage formed when the battery is connected to an external circuit), R_0 represents the internal ohmic resistance, R_1, and R_2 represent the electrochemical polarization resistance and concentration polarization resistance, and C_1 and C_2 represent the electrochemical polarization capacitance and concentration polarization capacitance, respectively. The first R_1C_1 parallel network represents the rapid voltage change process when the current changes in the chemical reaction process inside the battery, and the second R_2C_2 parallel network represents the slow voltage change process during the chemical reaction process inside the battery. U_1 is the terminal voltage of the first R_1C_1 parallel network circuit, and U_2 is the terminal voltage of the second R_2C_2 parallel network circuit. Assuming that the current I is in the positive direction during discharge, the voltage and current expression of the second-order RC equivalent model can be obtained from Kirchhoff's voltage and current law, as shown in Eq. (3-1).

$$\begin{cases} U_L = U_{OC} - IR_0 - U_1 - U_2 \\ \dfrac{dU_1}{dt} = \dfrac{I}{C_1} - \dfrac{U_1}{R_1C_1} \\ \dfrac{dU_2}{dt} = \dfrac{I}{C_2} - \dfrac{U_2}{R_2C_2} \end{cases} \tag{3-1}$$

In Eq. (3-1), U_{OC} is open-circuit voltage OCV, and the relation between U_{OC} and SOC can be established according to the polynomial of the OCV-SOC fitting curve. $x = [SOC, U_1, U_2]^T$ is selected as the state variable, and the equivalent circuit is discretized to obtain the discretized state space expression, as shown in Eq. (3-2).

$$\begin{cases} \begin{bmatrix} SOC_{k+1} \\ U_{1,k+1} \\ U_{2,k+1} \end{bmatrix} = \begin{bmatrix} 1 & 0 & 0 \\ 0 & e^{-\frac{\Delta t}{\tau_1}} & 0 \\ 0 & 0 & e^{-\frac{\Delta t}{\tau_2}} \end{bmatrix} \begin{bmatrix} SOC_k \\ U_{1,k} \\ U_{2,k} \end{bmatrix} + \begin{bmatrix} -\dfrac{\eta \Delta t}{Q_0} \\ R_1\left(1 - e^{-\frac{\Delta t}{\tau_1}}\right) \\ R_2\left(1 - e^{-\frac{\Delta t}{\tau_2}}\right) \end{bmatrix} I_k + w_k \\ U_{L,k+1} = U_{OC,k+1} + \begin{bmatrix} 0 & -1 & -1 \end{bmatrix} \begin{bmatrix} SOC_{k+1} \\ U_{1,k+1} \\ U_{2,k+1} \end{bmatrix} - IR_0 + v_k \end{cases} \tag{3-2}$$

In Eq. (3-2), the first expression is the state equation of the second-order model, and the second expression is its observation equation. Δt said sampling time interval, τ as the time constant, $\tau_1 = R_1C_1$, $\tau_2 = R_2C_2$. RC pairs with different time constants represent diffusion processes in electrolytes and porous electrodes as well as charge transfer and bilayer effects in electrodes. w_k and v_k are system state noise and measurement noise respectively. Q_0 means the capacity of the battery, η for coulomb efficiency.

3.2. Online parameter identification

Parameter estimation is a mathematical method to determine the parameters of a model by collecting data when the model structure is known (Du et al., 2023c; Wang et al., 2021a; Xiao et al., 2022). The recursive least square (RLS) algorithm is developed based on the theory of adaptive filtering (Shi et al., 2021a; Shi et al., 2022c; Shi et al., 2022d). The basic principle of the recursive least square algorithm is to estimate the coefficient on the premise of known input and output so that the sum of squares of the difference between the actual value and the estimated value is minimized (Kong et al., 2021a; Miao et al., 2021; Xia et al., 2022).

3.2.1. Limited memory recursive least square algorithm

The RLS method can achieve the adaptive identification of model parameters by constantly revising and updating the system parameters to accurately obtain the real-time characteristics of the system (Du et al., 2023b; Wang et al., 2022b; Zhang et al., 2023b). It has a wide range of applications in the field of system identification. The equation form of the parameters to be identified in the RLS principle is shown in Eq. (3-3).

$$y(k) = \phi(k)\theta(k)^T + e(k) \tag{3-3}$$

In Eq. (3-3), $y(k)$ is the output value of the system at time k, $\phi(k)$ as the observation vectors, $\theta(k)$ is the parameter vector to be estimated, and $e(k)$ is the noise observation vector.

The RLS algorithm is simple and practical, but with the increase of recursion times and data volume, a data saturation phenomenon will occur (He et al., 2021; Li et al., 2021a; Liu et al., 2021a). This is because the gain matrix gradually approaches zero with the increase of old data so that the correction ability of the recursive method slowly fails, and finally leads to the increasing error of parameter identification. To optimize the RLS algorithm, the forgetting factor λ is introduced to weaken the influence of old data and enhance the feedback effect of new data, namely, the forgetting factor recursive least square (FFRLS) algorithm (Long et al., 2022; Xie et al., 2021; Zhang et al., 2022d). Although FFRLS improves the shortcomings of RLS to some extent, there is still data supersaturation when there is too much data. Given the defects and deficiencies of the RLS algorithm and FFRLS algorithm, this book designed the limited memory least square method (LMRLS), that is, when the input of new data, removes the old data, only use the limited length of the latest data for parameter estimation (Cao et al., 2023; Kong et al., 2021b; Wang et al., 2023a; Wei et al., 2014). FFRLS are weighted to reduce the impact of "old" data based on how "old" it is. LMRLS algorithm, on the other hand, directly removes the old data and calculates the output at the current moment based on a limited set of data. It can obtain more information from the new data and has a better identification effect compared with the FFRLS algorithm. Therefore, the LMRLS algorithm is suitable for the estimation of time-varying parameters, which can effectively overcome data saturation and avoid estimation divergence.

3.2.2. Design the flow of the LMRLS algorithm

The memory length of the limited memory least square method is initialized as *L*, and its recursion principle is shown as follows.

Step 1: Set the initial value, and obtain the parameter estimation value of the memory interval by using RLS recursive algorithm. The recursive formula of RLS is shown in Eq. (3-4).

$$\begin{cases} \hat{\theta}(k+1) = \hat{\theta}(k) + K(k+1)[y(k+1) - \phi^T(k+1)\hat{\theta}(k)] \\ K(k+1) = P(k+1)\phi(k+1)[\phi^T(k+1)P(k)\phi(k+1) + 1]^{-1} \\ P(k+1) = [H - K(k+1)\phi^T(k+1)]P(k) \end{cases} \tag{3-4}$$

Step 2: The parameter estimation results obtained by the RLS algorithm are taken as the initial values of the limited memory least square method. The output at moment *k* is calculated based on the *L*+1 set of data corresponding to the moment *k-L* to moment k and the result (initial value) obtained by recursion at moment k-1, and the calculation process is shown in Eq. (3-5).

$$\begin{cases}\hat{\theta}(k-L,k)=\hat{\theta}(k-L,k-1) \\ \qquad +K(k-L,k)[y(k)-\phi^T(k)\hat{\theta}(k-L,k-1)] \\ K(k-L,k)=P(k-L,k-1) \\ \qquad \cdot\phi(k)[\phi^T(k)P(k-L,k-1)\phi(k)+1]^{-1} \\ P(k-L,k)=[H-K(k-L,k)\phi^T(k)]P(k-L,k-1)\end{cases} \tag{3-5}$$

Step 3: To keep the memory length L constant, a set of data at moment k is added, and a set of data at moment k-L needs to be removed. According to the L sets of data corresponding to the moment, k-L+1 to moment k and the results obtained from Eq. (3-5), the final solution formula of the limited memory recursive least squares method at moment k can be obtained as shown in Eq. (3-6).

$$\begin{cases}\hat{\theta}(k-L+1,k)=\hat{\theta}(k-L,k) \\ \qquad -K(k-L+1,k)[y(k-L)-\phi^T(k-L)\hat{\theta}(k-L,k)] \\ K(k-L+1,k)=P(k-L,k) \\ \qquad \cdot\phi(k-L)[\phi^T(k-L)P(k-L,k)\phi(k-L)+1]^{-1} \\ P(k-L+1,k)=[H+K(k-L+1,k)\phi^T(k-L)]P(k-L,k)\end{cases} \tag{3-6}$$

Where K (*) is the Kalman gain function, P (*) is the covariance function, and I is the unit matrix. From the above analysis, the recursive schematic flow chart of the LMRLS algorithm can be obtained as shown in Fig. 3-3.

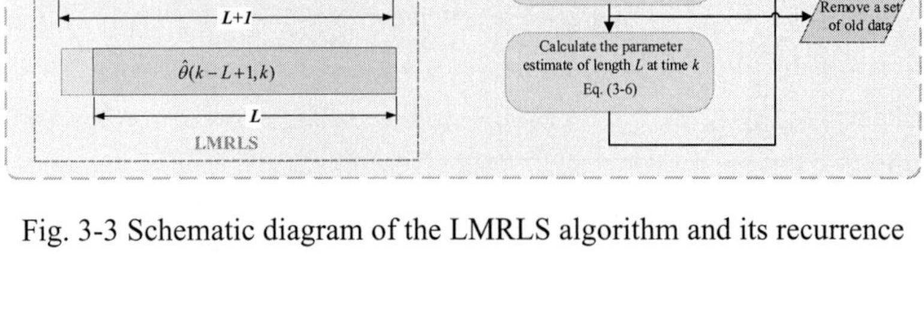

Fig. 3-3 Schematic diagram of the LMRLS algorithm and its recurrence

Where, $\hat{\theta}(k-L,k-1)$ represents the parameter identification result corresponding to the L set of data from the moment k-L to moment k-1, $\hat{\theta}(k-L,k)$ denotes the parameter identification result corresponding to the L+1 set of data from the moment k-L to moment k, $\hat{\theta}(k-L+1,k)$ stands for the parameter identification result corresponding to the L set of data from the moment k-L+1 to moment k.

3.2.3. Establishing the parameter identification model of LMRLS

According to the LMRLS algorithm design process in the previous section, the LMRLS parameter identification model for second-order RC equivalent circuits will be established in this section. According to Fig. 3-2 and Eq. (3-1), the mathematical relationship can be obtained by Kirchhoff's law and Laplace transform, as shown in Eq. (3-7).

$$U_L(s) = U_{OC}(s) - I(s)\left(R_0 + \frac{R_1}{R_1C_1s+1} + \frac{R_2}{R_2C_2s+1}\right) \tag{3-7}$$

Make

$$\begin{cases} \tau_1 = R_1C_1 \\ \tau_2 = R_2C_2 \end{cases} \tag{3-8}$$

$$\begin{cases} a = \tau_1\tau_2 \\ b = \tau_1 + \tau_2 \\ c = R_1 + R_2 + R_0 \\ d = R_1\tau_2 + R_2\tau_1 + R_0(\tau_1+\tau_2) \end{cases} \tag{3-9}$$

Substituting Eq. (3-8) and Eq. (3-9) into Eq. (3-7) in turn, we can get

$$aU_{OC}s^2 + bU_{OC}s + U_{OC} = aR_0{I_s}^2 + dI_s + cI + aU_s^2 + bU_s + U \tag{3-10}$$

Let

$$\begin{cases} s = \dfrac{x(k) - x(k-1)}{T} \\ s^2 = \dfrac{x(k) - 2x(k-1) + x(k-2)}{T^2} \end{cases} \tag{3-11}$$

Where T is the sampling time, and Eq. (3-11) is substituted into Eq. (3-10) to obtain.

$$\begin{aligned} U_{OC}(k) - U_L(k) = k_1[U_L(k-1) - U_{OC}(k-1)] \\ + k_2[U_L(k-2) - U_{OC}(k-2)] \\ + k_3 I(k) + k_4 I(k-1) + k_5 I(k-2) \end{aligned} \tag{3-12}$$

In Eq. (3-12),

$$\begin{cases} k_1 = \dfrac{-bT - 2a}{T^2 + bT + a} \\ k_2 = \dfrac{a}{T^2 + bT + a} \\ k_3 = \dfrac{cT^2 + dT + aR_0}{T^2 + bT + a} \\ k_4 = \dfrac{-dT - 2aR_0}{T^2 + bT + a} \\ k_5 = \dfrac{aR_0}{T^2 + bT + a} \end{cases} \tag{3-13}$$

Eq. (3-12) can be put into the RLS method, $\theta = (k_1, k_2, k_3, k_4, k_5)^T$ will be employed as the parameters of direct identification, and then the identification results of these parameters can be used to deduce the second-order RC equivalent circuit model parameters R_0, R_1, R_2, C_1, and C_2. The derivation is as follows.

The relationship $k_0 = T^2 + bT + a$ is obtained by Eq. (3-13) available for each parameter value, as shown in Eq. (3-14).

$$\begin{cases} k_0 = \dfrac{T^2}{k_1 + k_2 + 1} \\ a = k_0 k_2 \\ b = \dfrac{-k_0(k_1 + 2k_2)}{T} \\ c = \dfrac{k_0(k_3 + k_4 + k_5)}{T^2} \\ d = \dfrac{-k_0(k_4 + 2k_5)}{T} \\ R_0 = k_5/k_2 \end{cases} \tag{3-14}$$

Setting $a=\tau_1\tau_2$, $b=\tau_1+\tau_2$, $\tau^2-b\tau+a=0$, the time constant can be obtained for τ_1 and τ_2, as shown in Eq. (3-15).

$$\begin{cases}\tau_1=\dfrac{b+\sqrt{b^2-4a}}{2}\\[2mm]\tau_2=\dfrac{b-\sqrt{b^2-4a}}{2}\end{cases}\tag{3-15}$$

By substituting the expressions c and d in Eq. (3-9) into Eq. (3-15), the remaining parameter values in the second-order RC equivalent model can be obtained.

To sum up, each parameter value of the battery model obtained by the LMRLS algorithm is shown in Eq. (3-16), and its schematic diagram is shown in Fig. 3-4.

$$\begin{cases}R_0=k_5/k_2\\R_1=(\tau_1c+\tau_2R_0-d)/(\tau_1-\tau_2)\\R_2=c-R_1-R_0\\C_1=\tau_1/R_1\\C_2=\tau_2/R_2\end{cases}\tag{3-16}$$

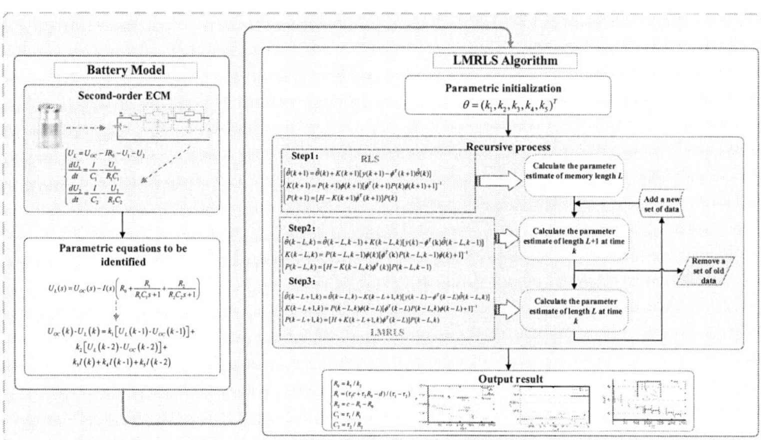

Fig. 3-4 Schematic diagram of the second-order LMRLS recursive flow chart

As shown in Fig. 3-4, it clearly shows the logical thinking of the whole parameter identification. Firstly, the parameter equation to be identified is obtained through ECM, and then the final identification result is output through LMRLS recursive operation, to prepare for the collaborative estimation in the next step.

3.3. Result verification and analysis of the LMRLS model

According to the experimental data in Section 2.2, complex working conditions at different temperatures, the collected experimental data are verified by parameter identification of the second-order model after the three experiments under HPPC, DST, and BBDST.

3.3.1. Online parameter identification results of LMRLS

Under HPPC conditions of different temperatures, the second-order model parameter identification results of the LMRLS algorithm are shown in Fig. 3-5.

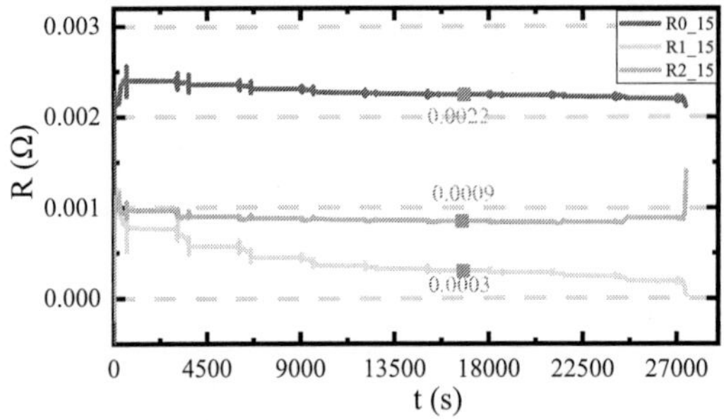

(a) Identification result of resistance at 15℃

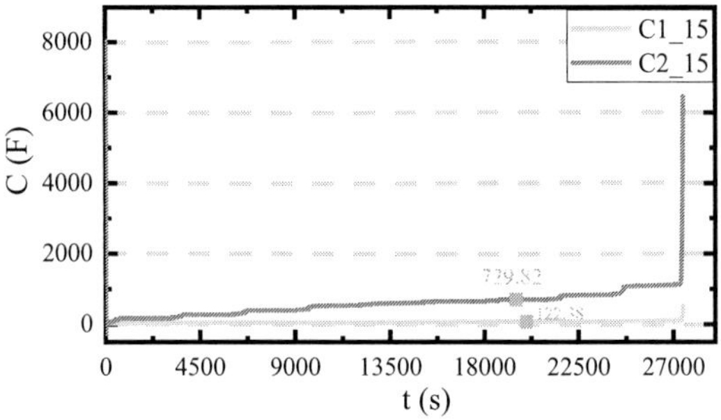

(b) Identification results of capacitance at 15℃

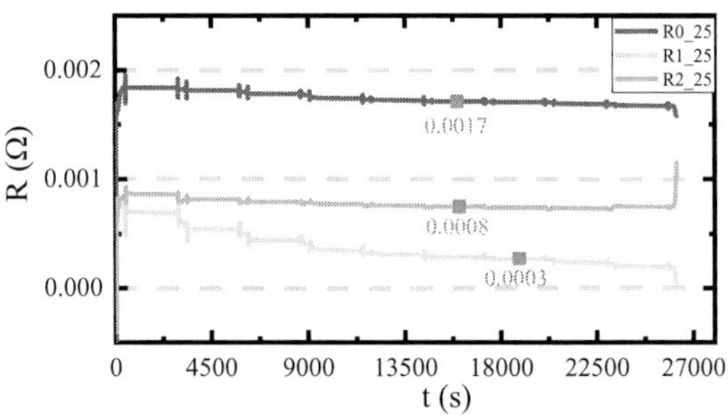

(c) Identification result of resistance at 25℃

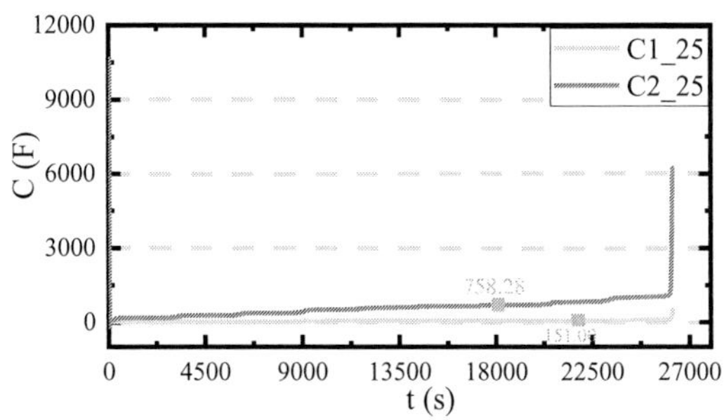

(d) Identification results of capacitance at 25℃

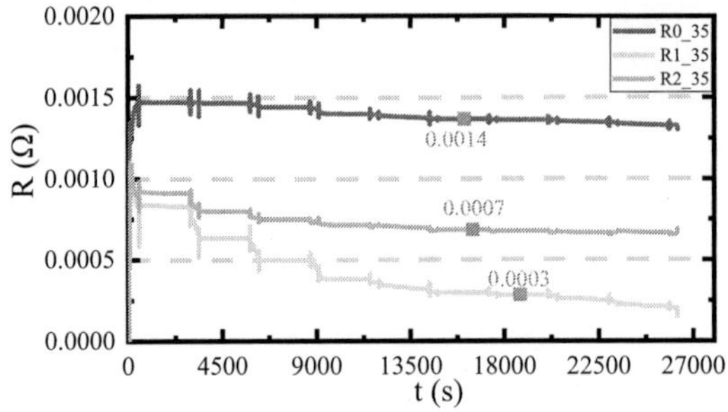

(e) Identification result of resistance at 35℃

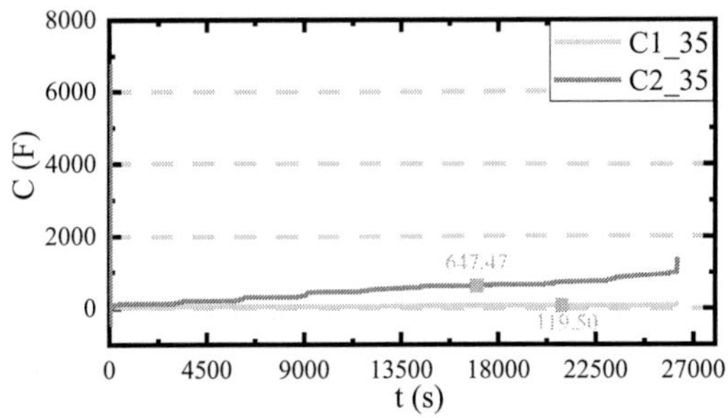

(f) Identification results of capacitance at 35℃

Fig. 3-5 Parameter identification results of HPPC working conditions at different temperatures

As can be seen from Fig. 3-5, the variation trend of resistance and capacitance at different temperatures is the same, but there are some differences in values. At the beginning and end of battery discharge, due to the chemical reaction inside the battery, the resistance value and capacitance value change greatly. In the discharge process, the resistance value and capacitance value gradually tend to be relatively stable.

3.3.2. Comparison of results of different parameter identification models

To verify the tracking capability of the parameter identification algorithm, the OCV-SOC function is applied to the RLS, FFRLS, and LMRLS algorithms respectively to obtain the analog voltage value. Then the simulated voltage is compared with the experimental voltage to verify the performance of the parameter identification algorithm. The evaluation indexes of experimental results in this study are represented by maximum error (MAXE), mean absolute error (MAE) and root mean square error (RMSE), respectively. Eq. (3-17) is the calculation formula of each evaluation index, Fig. 3-6 and Table 3-1 show the comparison of simulated voltage results of RLS, FFRLS, and LMRLS algorithms under HPPC working conditions at different temperatures.

$$\begin{cases} MAXE = max|x_k - \hat{x}_k| \\ MAE = \dfrac{1}{n}\sum_{i=1}^{n}|x_k - \hat{x}_k| \\ RMSE = \sqrt{\dfrac{1}{n}\sum_{i=1}^{n}(x_k - \hat{x}_k)^2} \end{cases} \tag{3-17}$$

Where x_k is reference value at time k, $\hat{x}_k$ is system estimation value at moment k, and n is the total number of samples.

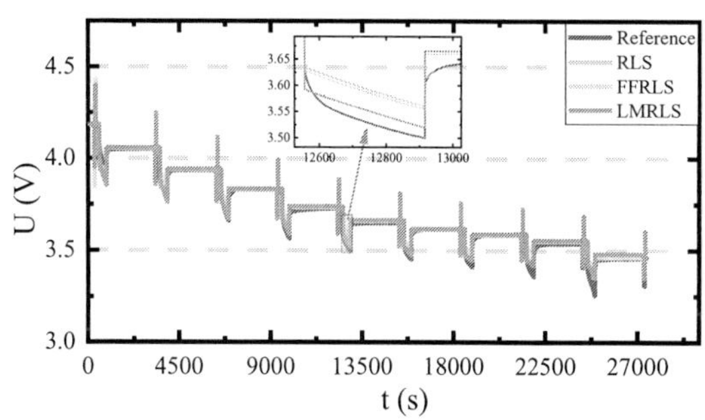

(a) Voltage simulation comparison of different algorithms at 15℃

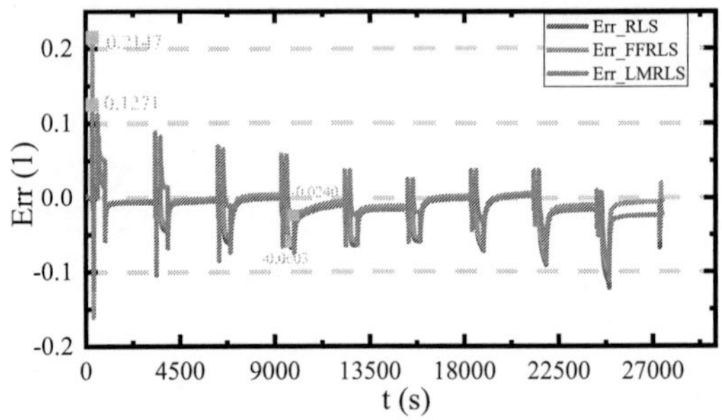

(b) Voltage simulation error of different algorithms at 15℃

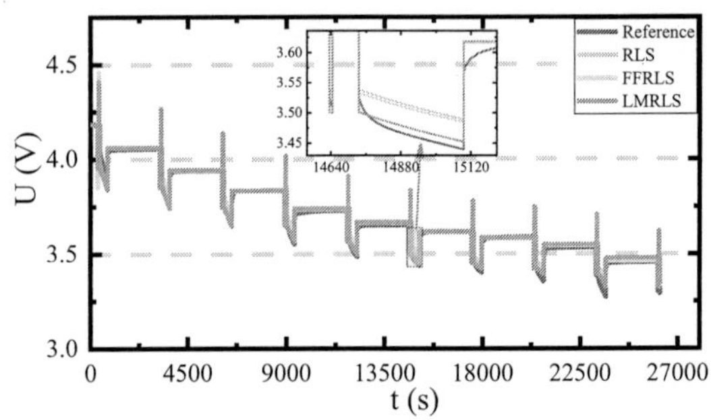

(c) Voltage simulation comparison of different algorithms at 25℃

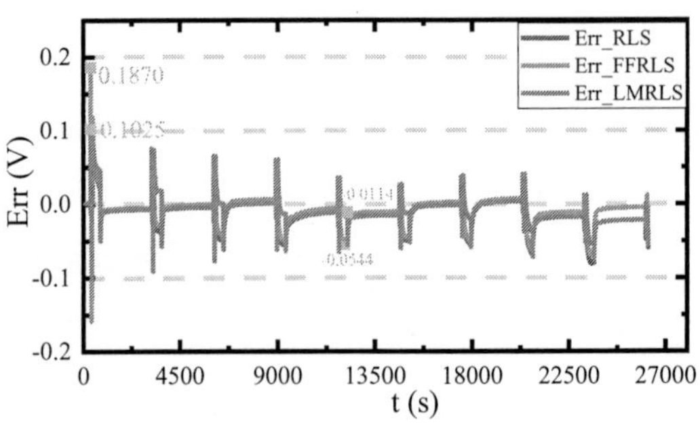

(d) Voltage simulation error of different algorithms at 25℃

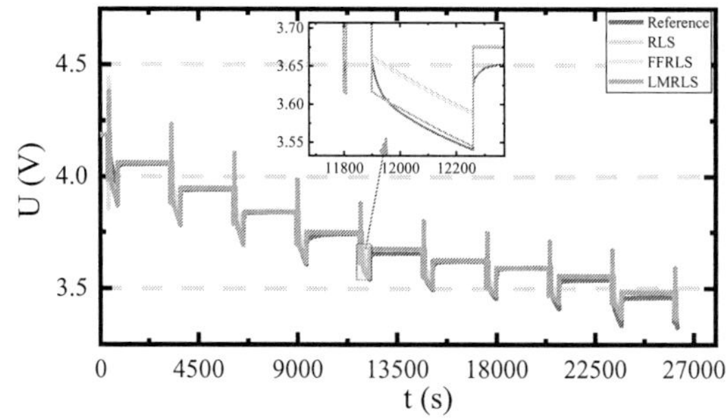

(e) Voltage simulation comparison of different algorithms at 35℃

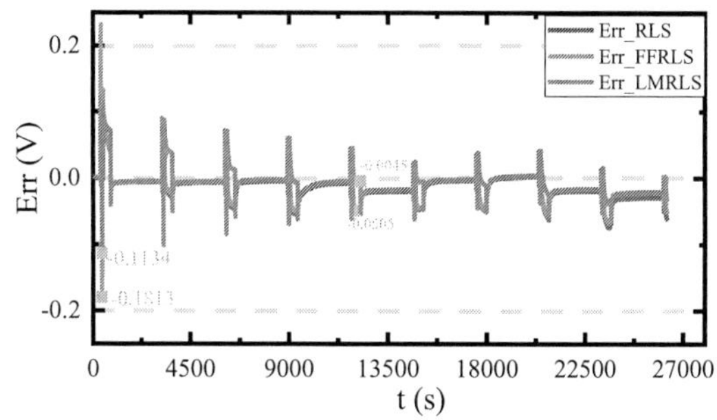

(f) Voltage simulation error of different algorithms at 35℃

Fig. 3-6 Voltage simulation results of different algorithms under HPPC working conditions with different temperatures

Table 3-1 Voltage simulation errors of different algorithms under HPPC conditions

(a) Voltage simulation errors at 15℃

Estimation method	MAXE	MAE	RMSE
RLS	54.76%	1.85%	3.37%
FFRLS	51.67%	1.41%	2.94%
LMRLS	12.61%	1.41%	2.01%

(b) Voltage simulation errors at 25℃

Estimation method	MAXE	MAE	RMSE
RLS	45.72%	1.73%	2.94%
FFRLS	41.94%	1.30%	2.45%
LMRLS	10.21%	1.28%	1.74%

(c) Voltage simulation errors at 35℃

Estimation method	**MAXE**	**MAE**	**RMSE**
RLS	23.17%	1.79%	2.52%
FFRLS	23.09%	1.48%	2.20%
LMRLS	11.61%	1.25%	1.80%

It can be seen from (a), (c), and (e) in Fig. 3-6 that the LMRLS algorithm can well follow the curve of reference voltage under HPPC conditions of different temperatures. Especially at the moment of voltage abrupt change, the tracking ability is obviously superior to the RLS algorithm and FFRLS algorithm. It can be seen from (b), (d), and (f) in Fig. 3-6 that the overall error curve of the LMRLS algorithm fluctuates less in the whole process and is closer to 0. Especially, the error curve of the LMRLS algorithm fluctuates less than that of the RLS algorithm and FFRLS algorithm at the beginning and end of the experiment. As can be seen from the analysis of Table 3-1, the maximum error of the LMRLS algorithm remains at about 11% under different temperatures, which is significantly improved compared with RLS and FFRLS algorithms, while FFRLS and RLS have little difference. MAE and RMSE of the LMRLS algorithm are still superior at different temperatures, and both remain within 2%. In addition, the errors of the three algorithms tend to decrease with the increase in temperature. In summary, it can be seen that the LMRLS algorithm has a good convergence effect and tracking performance at different temperatures, which proves that the LMRLS algorithm has high robustness and accuracy.

To further verify the tracking ability of the LMRLS algorithm, the simulation voltage comparison of parameter identification in BBDST working conditions is carried out at 15°C, 25°C, and 35°C respectively. The results are shown in Fig. 3-7 and Table 3-2.

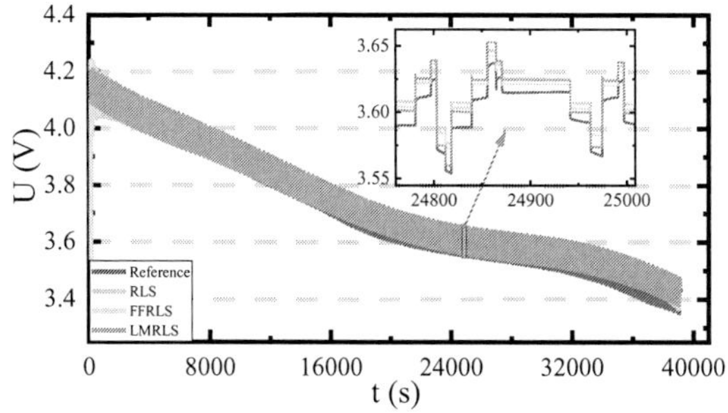

(a) Voltage simulation comparison of different algorithms at 15℃

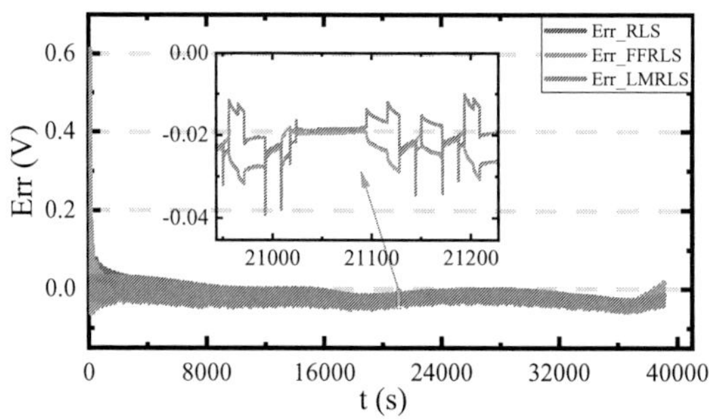

(b) Voltage simulation error of different algorithms at 15℃

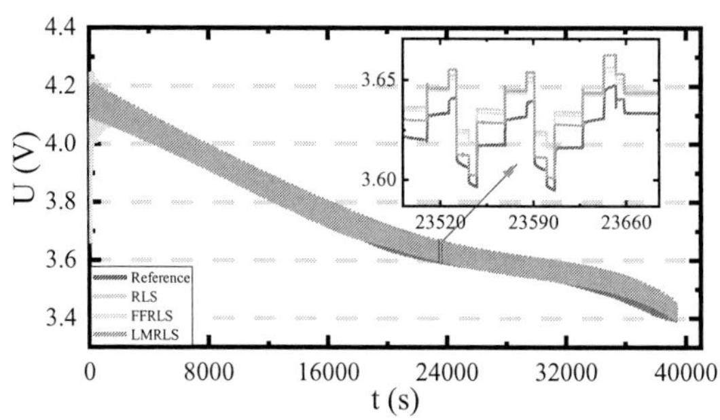

(c) Voltage simulation comparison of different algorithms at 25℃

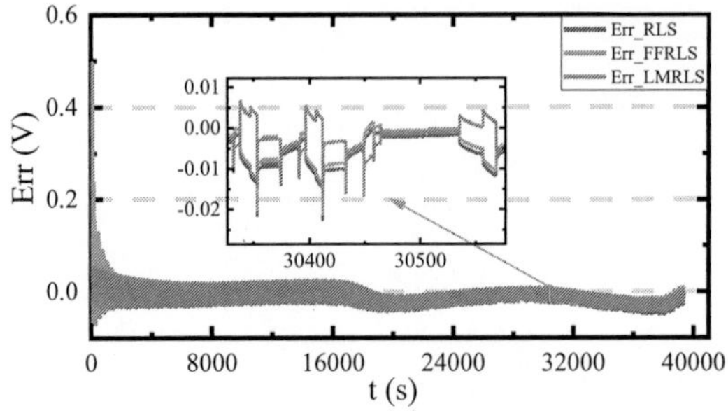

(d) Voltage simulation error of different algorithms at 25℃

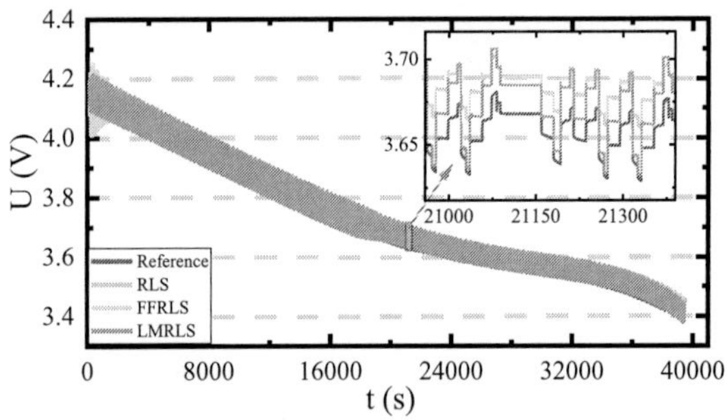

(e) Voltage simulation comparison of different algorithms at 35℃

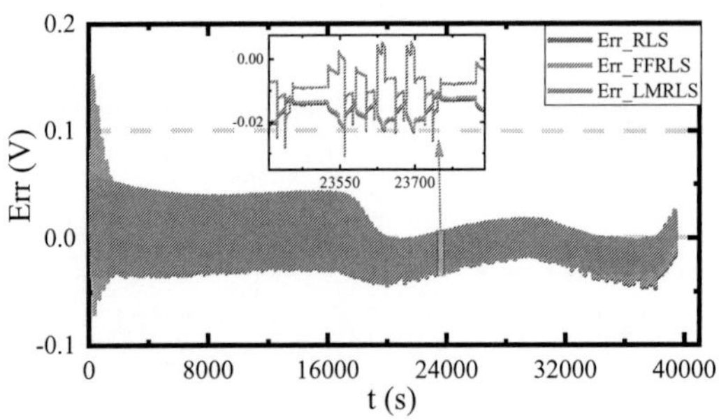

(f) Voltage simulation error of different algorithms at 35℃

Fig. 3-7 Voltage simulation results of different algorithms under BBDST working conditions with different temperatures

Table 3-2 Voltage simulation errors of different algorithms under BBDST conditions

(a) Voltage simulation errors at 15℃

Estimation method	MAXE	MAE	RMSE
RLS	17.02%	2.00%	2.23%
FFRLS	61.12%	1.57%	2.14%
LMRLS	5.89%	1.60%	1.82%

(b) Voltage simulation errors at 25℃

Estimation method	MAXE	MAE	RMSE
RLS	49.63%	1.22%	1.74%
FFRLS	49.61%	1.04%	1.59%
LMRLS	5.10%	1.01%	1.28%

(c) Voltage simulation errors at 35℃

Estimation method	MAXE	MAE	RMSE
RLS	15.17%	1.30%	1.67%
FFRLS	15.15%	1.23%	1.60%
LMRLS	5.95%	1.01%	1.36%

As can be seen from Fig. 3-7, the LMRLS algorithm has good tracking ability and convergence under different temperatures, with small overall error, while the RLS algorithm and FFRLS algorithm have poor tracking effect in the early stage, and gradually tend to be good in the middle and later stages. According to the further analysis in Table 3-2, the maximum error of the LMRLS algorithm is kept within 6%, and MAE and RMSE are kept within 2%. Compared with RLS and FFRLS algorithms, they all show excellent performance at different temperatures.

3.4. Chapter Summary

In this chapter, the advantages and disadvantages of different equivalent circuit models are analyzed first. After comparative analysis, the second-order equivalent circuit model is adopted to consider the accuracy and complexity of the model. Then because of the defects and deficiencies of the RLS algorithm and FFRLS algorithm, this book designed the LMRLS algorithm flow and built the LMRLS parameter identification model.

Finally, to verify the accuracy and convergence of the LMRLS parameter identification model, HPPC and BBDST conditions at different temperatures are verified. The experimental results show that the LMRLS algorithm has good tracking ability and robustness at different temperatures, and can accurately characterize the working state of power lithium-ion batteries, which provides a guarantee for the accurate state estimation of the lithium-ion battery.

CHAPTER 4

STRATEGY OF CORRECTING KALMAN FILTER BASED ON BPNN

SOC and SOH are important parameters to measure the battery status, which not only relate to whether the battery management system can effectively control the battery but also are important guarantees for the reliable power output of electric vehicles, improving the battery range and driving safety (Fan et al., 2021b; Xu et al., 2020; Zhang et al., 2020). At present, a large number of researches focus on SOC or SOH. Considering the coupling between SOC and SOH, it is difficult to guarantee the accuracy of the true state of a lithium battery by estimating a state separately. In this book, the collaborative estimation of SOC and SOH is realized by constructing a dual Kalman filter, and BPNN is introduced to compensate for the model error of the Kalman filter so that the accuracy and stability of the predicted results are further improved.

4.1. Analysis of SOC estimation strategy

The Kalman filter (KF) algorithm was proposed by the Hungarian mathematician Kalman in the 1960s and is one of the most widely used intelligent algorithms, often used in practical scenarios such as path planning, target tracking, and lithium battery SOC estimation (Chen et al., 2021b; Hossain et al., 2022; Urrea and Agramonte, 2021).

4.1.1. Extended Karman filtering

KF is a filtering theory created by the state space theory in the time domain, which is mainly used to estimate linear time-invariant systems (Jin et al., 2021b; Shi et al., 2021b; Wang et al., 2020a; Wang et al., 2020b). Its basic principle is that the state space model of signal and noise is constructed with the minimum mean square error as the best estimation criterion, and the relationship between the state variables and the observed variables is deduced from the state equation and the observed equation of the established model, and the estimated value of the previous moment and the observed value of the current moment are used to update the estimates of the state variables (Liu et al., 2020; Ma et al., 2022a; Wang et al., 2019a; Wang et al., 2019b). The core idea of Kalman filtering is to use the observable output estimation error of the system to repair the unobservable state estimation error and obtain the state estimation value corresponding to the minimum mean square error according to the principle of minimum mean square error in the finite number of iterative operations, so as to greatly reduce the noise interference in the data stream, make the estimated value is constantly approaching the real state value, thereby improving the estimation accuracy of the system (Khodaparast, 2022; Shi et al., 2020; Soltani et al., 2016; Wang et al., 2019c).

The essence of the Kalman filter algorithm to estimate the SOC of lithium batteries is to use the ampere-hour integration method to calculate SOC, and at the same time use the measured voltage value to correct the SOC value obtained by the ampere-hour integration method (Xiong et al., 2020). However, the Kalman filter algorithm itself is proposed for a linear state model, but in most cases, the working characteristics of lithium batteries are highly nonlinear, which requires the battery to be linearized forcibly to adapt to KF, that is, the extended Kalman filter (EKF) algorithm (Chen et al., 2021c; Mahony et al., 2022; Yu et al., 2022a). The EKF expands the nonlinear function of OCV and SOC around the SOC prediction into a Taylor series and omits the higher order terms above the second order so that an approximate linearized model can be obtained to transform the nonlinear problem into a linear one, which allows the Kalman linear filtering algorithm to be applied to nonlinear systems (Wang et al., 2020d).

When using EKF to estimate the SOC of a lithium battery, SOC is a component in the state vector, the current is used as the control variable in the input parameter, and the output is the terminal voltage calculated by the battery equivalent model. The system noise and observation noise are both Gaussian white noise, which variance denoted as Q and R. This method usually uses the state and the measurement equation of the system as the basis for derivation. The predicted state equation includes the ampere-hour integration method for calculating SOC, and the observation equation reflects the equivalent model of the lithium battery. The accuracy of using the EKF algorithm to estimate SOC depends largely on the accuracy of the equivalent model, so it is very important to establish a suitable equivalent model for lithium batteries. The principle and process of the basic EKF algorithm will be analyzed in detail below.

The state space equations of a discrete nonlinear system are shown in Eq. (4-1).

$$\begin{cases} X_{k+1} = f(X_k, k) + w_k \\ Z_k = h(X_k, k) + v_k \end{cases} \tag{4-1}$$

In Eq. (4-1), the first part represents the state equation and the second part denotes the observation equation. k is the discrete-time, X_{k+1} is the n-dimensional state vector, Z_k is the m-dimensional observation vector, and w_k and v_k are mutually independent Gaussian white noise. To apply the Kalman filter, a first-order Taylor expansion of the nonlinear functions $f(*)$ and $h(*)$ is expanded around $\hat{X}_k$. The expansion results are shown in Eq. (4-2).

$$\begin{cases} f(X_k,k) \approx f(\hat{X}_k,k) + \left.\dfrac{\partial f(X_k,k)}{\partial X_k}\right|_{X_k=\hat{X}_k} (X_k - \hat{X}_k) \\ h(X_k,k) \approx h(\hat{X}_k,k) + \left.\dfrac{\partial h(X_k,k)}{\partial X_k}\right|_{X_k=\hat{X}_k} (X_k - \hat{X}_k) \end{cases} \tag{4-2}$$

For Eq. (4-2), let

$$\begin{cases} A_k = \left.\dfrac{\partial f(X_k,k)}{\partial X_k}\right|_{X_k=\hat{X}_k} \\ B_k = f(\hat{X}_k,k) - A_k\hat{X}_k \\ C_k = \left.\dfrac{\partial h(X_k,k)}{\partial X_k}\right|_{X_k=\hat{X}_k} \\ D_k = h(\hat{X}_k,k) - C_k\hat{X}_k \end{cases} \tag{4-3}$$

Further, Eq. (4-1) can be linearized as shown in Eq. (4-4).

$$\begin{cases} X_{k+1} = A_kX_k + B_k + w_k \\ Z_k = C_kX_k + D_k + v_k \end{cases} \tag{4-4}$$

The recursive process of EKF obtained by applying the basic KF method to the linearized model is shown as follows.

Step 1: State prediction.

$$\hat{X}_{k+1}^- = f(\hat{X}_k) \tag{4-5}$$

Step 2: Prediction of covariance.

$$\hat{P}_{k+1}^- = A_k\hat{P}_kA_k^T + Q_{k+1} \tag{4-6}$$

Step 3: Calculating Kalman gain.

$$K_{k+1} = \hat{P}_{k+1}^-C_{k+1}^T\left(C_{k+1}\hat{P}_{k+1}^-C_{k+1}^T + R_{k+1}\right)^{-1} \tag{4-7}$$

Step 4: State update.

$$\hat{X}_{k+1} = X_{k+1}^- + K_{k+1}[Z_{k+1} - h(X_{k+1}^-)] \tag{4-8}$$

Step 5: Noise covariance update.

$$\hat{P}_{k+1} = [H - K_{k+1}C_{k+1}]P_{k+1}^- \tag{4-9}$$

Among them, *P* is the covariance and *K* is the Kalman gain. *H* is an n×m unit matrix. Q and R are the expectations of w and v respectively, which generally do not vary with the system. The initial filtering value and the filtering covariance are $\hat{X}_0 = E(X_0)$, $P_0 = E[(X_0 - \hat{X}_0)(X_0 - \hat{X}_0)^T]$, respectively.

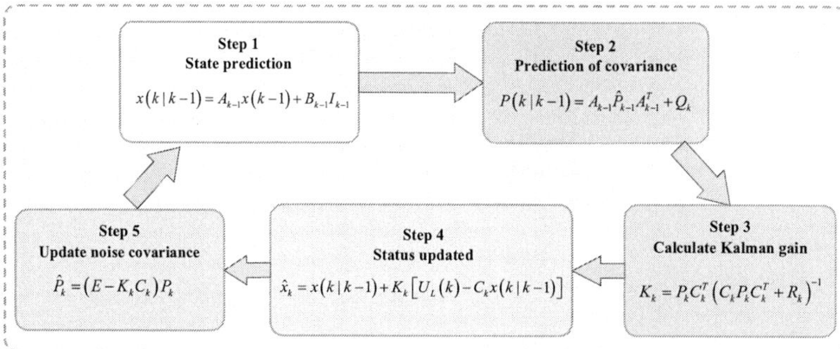

Fig. 4-1 The iterative calculation process of SOC estimated by the EKF

Fig. 4-1 illustrates the iterative computation flow of the EKF algorithm for predicting SOC. Where $x(k)$ is the state variable; A_k is the state transfer matrix; B_k is the system control input matrix; I_k is the input current value; C_k is the systematic observation matrix; K_k is the Kalman gain; P_k is the covariance matrix; Q_k is the expectation of the system process noise; R_k is the expectation of the system observation noise; $U_L(k)$ is the real-time terminal voltage value; and H is the unit matrix.

EKF continuously corrects the current estimated value through the Kalman gain, and can quickly track the true value of SOC during the continuous iterative operation process, and then obtain the optimal estimated value in the sense of the minimum mean square error. The algorithm has fast convergence speed and high estimation accuracy, can correct the initial error in battery estimation, and has a certain inhibitory effect on interference noise (Fan et al., 2021a; Li et al., 2022a; Wang et al., 2022a).

4.1.2. BPNN prognostic SOC/SOH model

The neural network has a strong nonlinear processing ability and self-learning ability, especially suitable for dealing with the nonlinear internal structure of lithium batteries (Chemali et al., 2018; Jiao et al., 2020; Xu et al., 2022). Because the information is distributed and stored in the neurons of the network, the fault tolerance and robustness are high (Li et al., 2021c; Takyi-Aninakwa et al., 2022c; Wang et al., 2022l). As shown in Fig. 4-2, there is a schematic diagram of a typical neural network model.

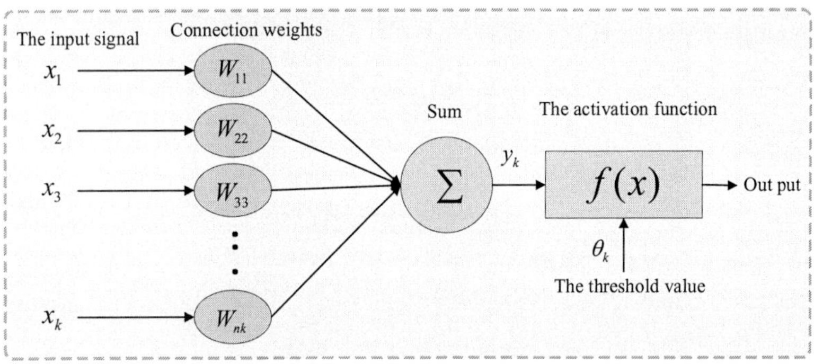

Fig. 4-2 Schematic diagram of a typical neural network model

As can be seen from Fig. 4-2, the signals are summed after being weighted with different weights and then processed by the activation function to output the results. The corresponding calculation principle of the neural network is shown in Eq. (4-10).

$$\begin{cases} y = \sigma(wx + b) \\ b_k \theta_j j W_{ik} ik V_{kj} \\ h_k = \sigma(\sum_{i=1}^{m} W_{ik} x_i + b_k) \\ y_j = \sigma(\sum_{k=1}^{l} V_{kj} h_k + \theta_j) \\ (x, y)(E) \\ \bar{y} = (\bar{y}_1, \bar{y}_2, \cdots, \bar{y}_n) \\ E = \frac{1}{2} \sum_{j=1}^{n} (\bar{y}_j - y_j)^2 \end{cases} \quad (4\text{-}10)$$

Where, the input quantity corresponding to x is expressed as $x = [x_0, x_1, x_2, \cdots, x_n]$, the weight value connected to it is $w = [w_0, w_1, w_2, \cdots, w_n]$, the corresponding threshold is b, and the final activation function is $\sigma(x)$. The output is y.

The BPNN is a multi-layer feed-forward neural network. The main feature of this network is the forward transmission of signals and the backward propagation of errors, consisting of an input layer, a hidden layer, and an output layer (Bao et al., 2022b; Li et al., 2022b; Lin et al., 2022). The main role of the hidden layer is to deal with nonlinear problems, which can be a relevant set according to the complexity of the problem being dealt with (Chen et al., 2021d; Du et al., 2022; Zhang et al., 2023a). When BPNN is trained, the first step is the forward propagation process, where the input data signal enters the neural network from the input layer and is processed layer by layer through the hidden layer until the output layer. If the output layer does not satisfy the condition of the set desired output, it enters the second step of the back propagation process, which adjusts the weights and thresholds of the network according to the prediction error, so that the predicted output of the BPNN continuously approximates the result of the desired output. Since the internal of lithium battery is a nonlinear system, it is particularly suitable for BPNN processing. The model structure of the BP neural network to predict SOC and SOH is shown in Fig. 4-3.

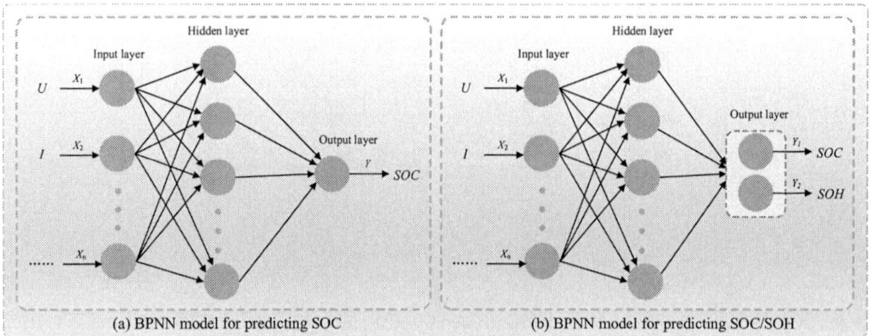

Fig. 4-3 BPNN model for SOC and SOH estimation

Fig. 4-3 (a) shows the network model for the BPNN estimation of SOC and Fig. 4-3 (b) shows the network model for the BPNN estimation of SOC and SOH. From Fig. 4-3, we can intuitively visualize the logical relationship between the input layer, hidden layer, and output layer of the BPNN. The parameters of the input layer are voltage, current, temperature, internal resistance, and other parameters related to SOC and SOH estimation accuracy. The number of nodes in the hidden layer can be set according to the complexity of the processing problem and the number of input parameters. The output of the layer is the predicted target SOC and SOH of the neural network.

4.1.3. SOC estimation based on BPNN correction strategy

The EKF algorithm converges quickly and the estimation results are stable, but it neglects the higher-order terms of the Taylor expansion when dealing with nonlinear systems, so there is a certain amount of model error. The model error will indirectly lead to a decrease in the accuracy of the estimation results. To compensate for the model error of EKF, BPNN is introduced. With the powerful self-learning ability and nonlinear processing capability of the neural network, the error curve function is obtained by learning and training the relevant parameters affecting the filter estimates until the error is less than the preset value. The obtained error curve function is then used to correct the estimation results of EKF to further improve the accuracy and stability of the results. The principle flow of the BPNN-EKF algorithm is shown in Fig. 4-4.

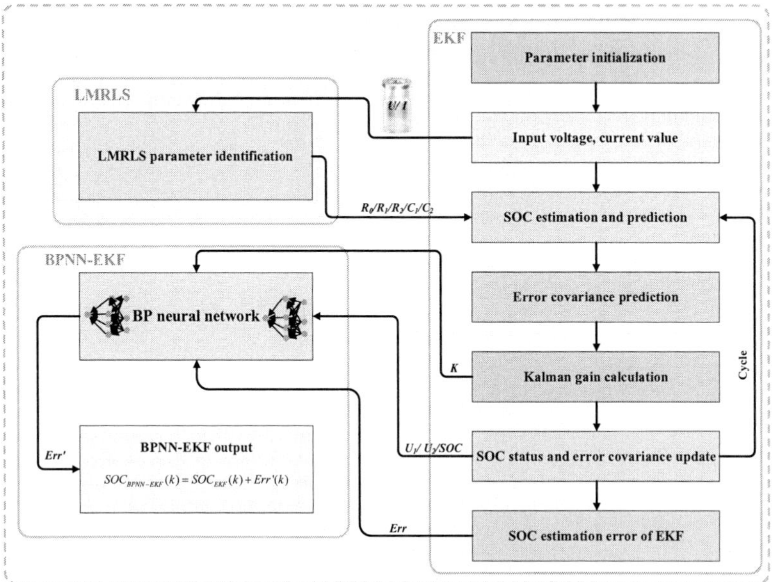

Fig. 4-4 General flow of the BPNN-EKF algorithm for estimating SOC

As can be seen from Fig. 4-4, the whole SOC estimation process consists of three modules: the LMRLS online parameter identification module, the EKF online state estimation module, and the BPNN error correction module. The latter two of these modules form the BPNN-EKF algorithm, which is the core of SOC estimation.

4.2 Constructing the dual extended Kalman co-estimation

There is a coupling effect between SOC and SOH, and it is difficult to guarantee the accurate estimation of the real state of the battery by estimating one state alone, so this study constructs a dual extended Kalman filter to achieve the cooperative estimation of SOC and SOH (Qiao et al.,

2023; Wang et al., 2020c).

4.2.1. SOC/SOH co-estimation strategy

Where, the first EKF realizes the estimation of SOC, and uses the estimation result as the input of the second EKF to realize the progressive estimation of SOH. Then, the estimated SOH value at the current time obtained by the second EKF is used as the input of the next iteration cycle to further correct the estimated SOC value at the next time, forming a closed loop until the end of the iteration. After the mutual correction feedback of the two, the co-estimation of the two is finally realized. The collaborative prognostic process of the DEKF algorithm is shown in Fig. 4-5.

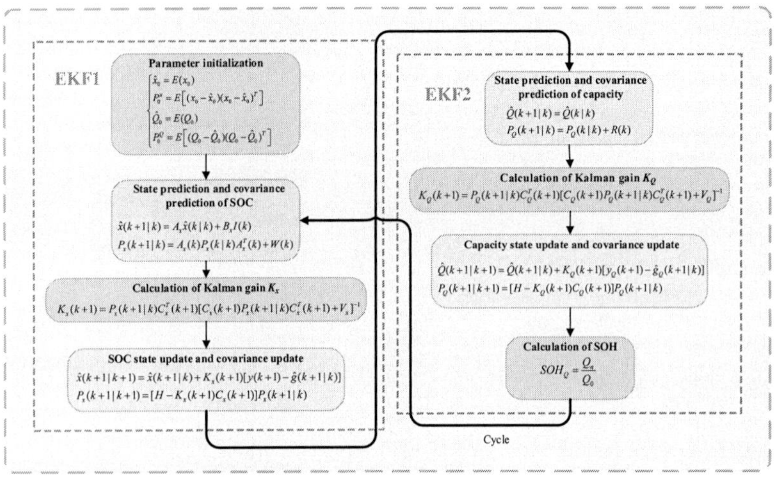

Fig. 4-5 Schematic diagram of DEKF synergistic estimation process

Fig. 4-5 clearly and intuitively visualizes the strategy process of dual Kalman collaborative estimation, which lays the foundation for the overall establishment of the collaborative prediction algorithm model in this study next.

4.2.2. Establishing ampere-hour model and capacity model

According to the second-order RC equivalent circuit model in Chapter 3, establish the ampere-hour model and capacity model of the battery, as shown in Table 4-1.

Table 4-1 Ampere-hour model and capacity model of lithium battery

Ampere-hour model	
$SOC_t = SOC_0 - \frac{\eta It}{Q_0}$	(4-11)
$\begin{cases} \begin{bmatrix} SOC_{k+1} \\ U_{1,k+1} \\ U_{2,k+1} \end{bmatrix} = \begin{bmatrix} 1 & 0 & 0 \\ 0 & e^{-\frac{\Delta t}{\tau_1}} & 0 \\ 0 & 0 & e^{-\frac{\Delta t}{\tau_2}} \end{bmatrix} \begin{bmatrix} SOC_k \\ U_{1,k} \\ U_{2,k} \end{bmatrix} + \begin{bmatrix} -\frac{\eta \Delta t}{Q_0} \\ R_1\left(1 - e^{-\frac{\Delta t}{\tau_1}}\right) \\ R_2\left(1 - e^{-\frac{\Delta t}{\tau_2}}\right) \end{bmatrix} I_k + w_k \\ U_{L,k+1} = U_{OC,k+1} - U_{1,k+1} - U_{2,k+1} - IR_0 + v_k \end{cases}$	(4-12)
Battery capacity model	
$SOH_Q = \frac{Q_n}{Q_0}$	(4-13)
$\begin{cases} Q_{k+1} = Q_k + r \\ d_k = SOC_k - SOC_{k-1} + \frac{\eta I_{t,k-1} T}{Q_{k-1}} + \xi \end{cases}$	(4-14)

In Eq. (4-11), SOC_t is the SOC value at moment t in the continuous moment, SOC_0 is the initial value of SOC, Q_0 is the rated capacity of the cell, η is the Coulomb efficiency, I is the current value, and t is the time.

In Eq. (4-12), the first expression and the second expression are the state equation and observation equation of SOC estimation respectively. The SOC_k is the SOC value at moment *k* after discretization of sampling of SOC_t, $U_{1,k}$ is the U_1 value at moment *k*, $U_{2,k}$ is the U_2 value at moment *k*, Δt is the sampling time interval, τ is the time constant, w_k is the system noise parameter, v_k is the observed noise parameter; I_k is the current value at moment *k*, $U_{L,k+1}$ is the terminal voltage value at moment *k*+1, and $U_{OC,k+1}$ is the open-circuit voltage value at moment *k*+1.

In Eq. (4-13), SOH_Q is the SOH value of the capacity model and Q_n is the current available capacity of the battery.

In Eq. (4-14), the first expression and the second expression, respectively, are the state equation and the observation equation for capacity estimation, d_k represents the difference between SOC at the current moment and SOC at the previous moment, r is the system noise, ξ is the measurement noise, and Q_k is the capacity value at moment k.

4.2.3. Design SOC/SOH co-estimation process

Based on the ampere-hour model and capacity model of the battery, the dual extended Kalman filter method is used to construct EKF1 and EKF2 to realize the collaborative estimation of SOC and SOH of the battery. Specific steps are as follows:

Step 1: Based on Eq. (4-12), with SOC as the only variable in the state equation, U_L as the variable in the observation equation, Q_0 as a constant, and the EKF1 is used to output the estimated value of SOC.

Step 2: Based on Eq. (4-14), take the estimated SOC value output by EKF1 as input, Q_n as the state variable, the difference d_k between the estimated SOC value at the current moment and the previous moment as the variable in the observation equation, and the EKF2 is used to output the estimated value of Q_n.

Step 3: Iteratively update the SOC estimates and Q_n estimates with EKF1 and EKF2, in which the SOC estimate output by EKF1 is taken as the input of EKF2 to obtain the Q_n estimate, and then calculate the SOH estimate by Eq. (4-13).

Step 4: Replace Q_0 in EKF1 with the estimated value of Q_n at the current moment, output the estimated value of SOC at the next moment, and get the estimated value of Q_n and SOH at the next moment according to Step 2 and Step 3, and iterate the loop accordingly to achieve the cooperative estimation of SOC and SOH.

The detailed process of DEKF construction is as follows:

EKF1

(a1) Parameter initialization

$$\begin{cases} \hat{x}_0 = E(x_0) \\ P_0^x = E[(x_0 - \hat{x}_0)(x_0 - \hat{x}_0)^T] \\ \hat{Q}_0 = E(Q_0) \\ P_0^Q = E\left[(Q_0 - \hat{Q}_0)(Q_0 - \hat{Q}_0)^T\right] \end{cases} \tag{4-15}$$

In Eq. (4-15), x_0 is the initial value of the state variable in EKF1, $\hat{x}_0$ is the expectation of x_0, P_0^x is the initial value of the covariance in EKF1, Q_0 is the initial value of the state variable in EKF2, and P_0^Q is the initial value of the covariance in EKF2.

(a2) According to Eq. (4-12), the state equation and observation equation of SOC are simplified as follows.

$$\begin{aligned} x_{k+1} &= A_x x_k + B_x I_k + w_k \\ y_{k+1} &= C_x x_{k+1} + D_x I_{k+1} + v_k \end{aligned} \tag{4-16}$$

In Eq. (4-16), x_k and x_{k+1} denote the state variables at moment k and k+1, respectively, A_x, B_x are the coefficient matrices of the state equations estimated by SOC, y_{k+1} is the observation at moment k+1, C_x, D_x are the coefficient matrices of the observation equations in EKF1, and I_{k+1} is the current at moment k+1.

(a3) State prediction and covariance prediction of SOC in EKF1

$$\begin{aligned} \hat{x}(k+1|k) &= A_x \hat{x}(k|k) + B_x I(k) \\ P_x(k+1|k) &= A_x(k) P_x(k|k) A_x^T(k) + W(k) \end{aligned} \tag{4-17}$$

In Eq. (4-17), $\hat{x}(k+1|k)$ is the preliminary estimated value of the state variable at time k+1 in EKF1, $\hat{x}(k|k)$ is the estimated value of the state variable at time k, and $P_x(k+1|k)$ is the error covariance matrix at time k+1, $P_x(k|k)$ is the error covariance matrix at time k, and $W(k)$ is the expectation of w_k.

(a4) Calculate Kalman gain *Kx*

$$K_x(k+1) = P_x(k+1|k)C_x^T(k+1)[C_x(k+1)P_x(k+1|k)C_x^T(k+1) + V_x]^{-1} \quad (4\text{-}18)$$

In Eq. (4-18), $K_x(k+1)$ is the Kalman gain at time *k*+1 in EKF1, and V_x is the expectation of v_k.

(a5) SOC state update and covariance update

$$\begin{aligned} \hat{x}(k+1|k+1) &= \hat{x}(k+1|k) + K_x(k+1)[y(k+1) - \hat{g}(k+1|k)] \\ P_x(k+1|k+1) &= [H - K_x(k+1)C_x(k+1)]P_x(k+1|k) \end{aligned} \quad (4\text{-}19)$$

In Eq. (4-19), $y(k+1)$ is the real-time observation at moment *k*+1 in EKF1, $\hat{g}(k+1|k)$ is the estimated observation at moment *k*+1, $\hat{x}(k+1|k+1)$ is the updated value of the state variable at moment *k*+1 in EKF1, $P_x(k+1|k+1)$ is the updated value of the covariance at moment *k*+1 in EKF1, and H denotes the unit matrix.

EKF2

(b1) According to Eq. (4-14), the state equation and observation equation about capacity in EKF2 are abbreviated as follows.

$$\begin{aligned} q_{k+1} &= q_k + r \\ d_{k+1} &= C_Q(x_{k+1} - x_k) + D_Q I_k + \xi \end{aligned} \quad (4\text{-}20)$$

In Eq. (4-20), q_k and q_{k+1} are the state variables at moment *k* and moment *k*+1, respectively, x_k and x_{k+1} are the state variables at moment *k*+1 and moment *k* in EKF1, respectively, and C_Q, D_Q are the coefficient matrices of the observation equations in EKF2.

(b2) State prediction and covariance prediction of capacity in EKF2

$$\begin{aligned} \hat{Q}(k+1|k) &= \hat{Q}(k|k) \\ P_Q(k+1|k) &= P_Q(k|k) + R(k) \end{aligned} \quad (4\text{-}21)$$

In Eq. (4-21), $\hat{Q}(k+1|k)$ is the preliminary estimated value of the state variable at time *k*+1 in EKF2, $\hat{Q}(k|k)$ is the estimated value of the state variable at time *k*, and $P_Q(k+1|k)$ is the error covariance matrix at time *k*+1 in EKF2, $P_Q(k|k)$ is the error covariance matrix at time *k*, and

$R(k)$ is the expectation of r.

(b3) Calculate the Kalman gain K_Q

$$K_Q(k+1) = P_Q(k+1|k)C_Q^T(k+1)[C_Q(k+1)P_Q(k+1|k)C_Q^T(k+1) + V_Q]^{-1} \quad (4\text{-}22)$$

In Eq. (4-22), $K_Q(k+1)$ is the Kalman gain at moment k+1 in EKF2 and V_Q is the expectation of ξ.

(b4) Capacity update and covariance update

$$\begin{aligned} \hat{Q}(k+1|k+1) &= \hat{Q}(k+1|k) + K_Q(k+1)[y_Q(k+1) - \hat{g}_Q(k+1|k)] \\ P_Q(k+1|k+1) &= [H - K_Q(k+1)C_Q(k+1)]P_Q(k+1|k) \end{aligned} \quad (4\text{-}23)$$

In Eq. (4-23), $y_Q(k+1)$ is the real-time observation at moment k+1 in EKF2, $\hat{g}_Q(k+1|k)$ is the estimated observation at moment k+1, $\hat{Q}(k+1|k+1)$ denotes the updated value of the state variable at moment k+1 in EKF2, and $P_Q(k+1|k+1)$ denotes the updated value of the covariance at moment k+1 in EKF2.

To sum up, the main process of the DEKF synergistic estimation algorithm is shown in Table 4-2.

Table 4-2 DEKF synergistic estimation algorithm flow

EKF1
Step 1: Parameter initialization
$\begin{cases} \hat{x}_0 = E(x_0) \\ P_0^x = E[(x_0 - \hat{x}_0)(x_0 - \hat{x}_0)^T] \\ \hat{Q}_0 = E(Q_0) \\ P_0^Q = E[(Q_0 - \hat{Q}_0)(Q_0 - \hat{Q}_0)^T] \end{cases}$
Step 2: State prediction and covariance prediction of SOC
$\hat{x}(k+1\|k) = A_x\hat{x}(k\|k) + B_xI(k)$ $P_x(k+1\|k) = A_x(k)P_x(k\|k)A_x^T(k) + W(k)$
Step 3: Calculate Kalman gain K_x
$K_x(k+1) = P_x(k+1\|k)C_x^T(k+1)[C_x(k+1)P_x(k+1\|k)C_x^T(k+1) + V_x]^{-1}$
Step 4: SOC state update and covariance update
$\hat{x}(k+1\|k+1) = \hat{x}(k+1\|k) + K_x(k+1)[y(k+1) - \hat{g}(k+1\|k)]$ $P_x(k+1\|k+1) = [H - K_x(k+1)C_x(k+1)]P_x(k+1\|k)$
EKF2
Step 5: State prediction and covariance prediction of capacity
$\hat{Q}(k+1\|k) = \hat{Q}(k\|k)$ $P_Q(k+1\|k) = P_Q(k\|k) + R(k)$

Step 6: Calculation of Kalman gain K_Q
$K_Q(k+1) = P_Q(k+1\|k)C_Q^T(k+1)[C_Q(k+1)P_Q(k+1\|k)C_Q^T(k+1) + V_Q]^{-1}$
Step 7: Capacity state update and covariance update
$\hat{Q}(k+1\|k+1) = \hat{Q}(k+1\|k) + K_Q(k+1)[y_Q(k+1) - \hat{g}_Q(k+1\|k)]$ $P_Q(k+1\|k+1) = [H - K_Q(k+1)C_Q(k+1)]P_Q(k+1\|k)$

4.3. SOC/SOH co-estimation based on the LMRLS-BPNN-DEKF model

4.3.1. BPNN modified the DEKF scheme

According to the analysis and study in Sections 4.1 and 4.2, the specific methodological flow of the BPNN to correct the DEKF is as follows:

Step 1: Obtain the parameters related to *SOC* and *capacity* at each moment in EKF1 and EKF2 to form the parameter matrix $S = [U_1, U_2, K_x, SOC, K_Q, Q]$, where K_x is the Kalman gain when SOC is estimated in EKF1, and K_Q is the Kalman gain when Q_n is estimated in EKF2. *SOC* and *Q* are the SOC estimated value and Qn estimated value output by EKF1 and EKF2 respectively. The error Err_{SOC} between the SOC estimated value and the reference value and the error Err_{SOH} between the SOH estimated value and the reference value are calculated for each moment to form the error matrix $Err = [Err_{SOC}, Err_{SOH}]$.

Step 2: The parameter matrix and error matrix are fed into the BP neural network for training until the errors Err_{SOC} and Err_{SOH} are less than the preset values, and the error curve functions Err_{SOC}' and Err_{SOH}' are obtained.

Step 3: The SOC estimate and SOH estimate are compensated by the error curve function, and the compensation formula is shown in Eq. (4-24).

$$\begin{cases} SOC_{BP-DEKF}(k) = SOC_{DEKF}(k) + Err_{SOC}'(k) \\ SOH_{BP-DEKF}(k) = SOH_{DEKF}(k) + Err_{SOH}'(k) \end{cases} \tag{4-24}$$

In Eq. (4-24), $SOC_{BP-DEKF}(k)$ and $SOH_{BP-DEKF}(k)$ are the SOC correction and SOH correction values at moment *k*, respectively, $SOC_{DEKF}(k)$ and $SOH_{DEKF}(k)$ are the SOC estimate and SOH estimate obtained by the dual extended Kalman filter method at moment k, respectively, and $Err_{SOC}'(k)$ and $Err_{SOH}'(k)$ are the values at moment k in Err_{SOC}' and Err_{SOH}', respectively.

4.3.2. LMRLS-BPNN-DEKF co-estimation algorithm model

Through the above analysis, the LMRLS-BPNN-DEKF collaborative estimation algorithm model can be obtained as shown in Fig. 4-6.

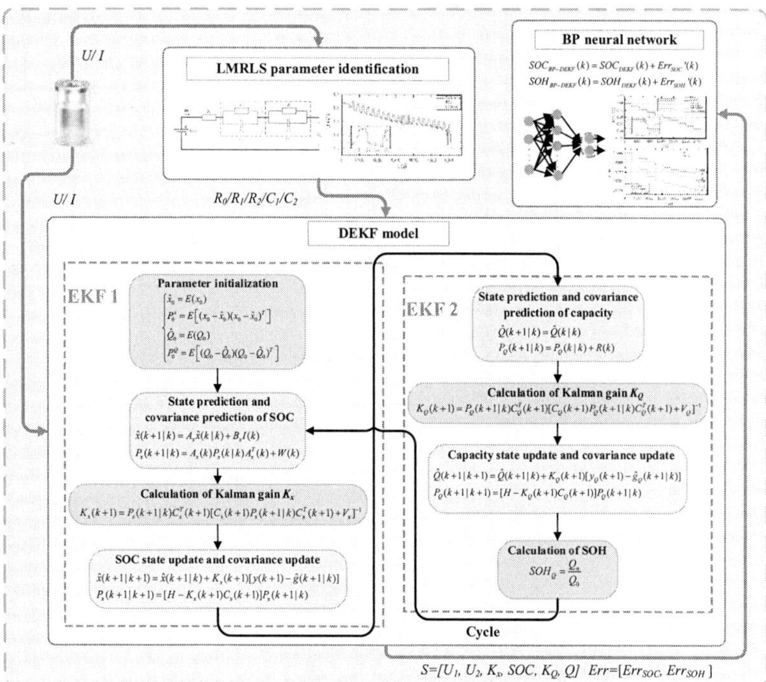

Fig. 4-6 Total flow of LMRLS-BPNN-DEKF collaborative estimation algorithm model

As can be seen from Fig. 4-6, firstly, the voltage and current data are input to the LMRLS parameter identification module and DEKF module respectively, and after LMRLS parameter identification, the results $[R_0, R_1, R_2, C_1, C_2]$ are transferred into the DEKF module for the SOC and

SOH co-estimation. Then the estimated parameter matrices $S = [U_1, U_2, K_x, SOC, K_Q, Q]$,$Err = [Err_{SOC}, Err_{SOH}]$ are passed into BPNN for training to obtain error curve functions $Err_{SOC}{}'$ and $Err_{SOH}{}'$, and finally, the final correction results of SOC and SOH are obtained by Eq. (4-23)

4.4. Chapter summary

This chapter mainly analyzes the method and strategy of the BP neural network correction Kalman filter and elaborates on the construction process of DEKF and the way of BPNN correction. Firstly, the ampere-hour model and capacity model of the battery are established based on the second-order RC equivalent circuit. Then a dual extended Kalman method is applied to estimate SOC and capacity of the battery by two EKFs respectively, and SOH is estimated by using the ampere-hour integral as a bridge, and the estimated values of SOC and SOH are correlated to form a closed loop, and the two correct each other with feedback to realize the co-estimation. Finally, a BPNN is introduced for correction, and the final SOC/SOH synergistic estimation based on the LMRLS-BPNN-DEKF model is realized.

CHAPTER 5

EXPERIMENTAL VERIFICATION AND ANALYSIS OF CO-ESTIMATION MODEL

To explore and study the accuracy and robustness of the LMRLS-BPNN-DEKF collaborative estimation model proposed in this study at different temperatures, an experimental platform is established in Chapter 2.2 of this book, and the HPPC, DST, and BBDST working conditions are tested at 15°C, 25°C and 35°C for a 70Ah ternary lithium-ion battery, respectively, which provides the necessary preparations for fully verifying the effectiveness and feasibility of the synergistic estimation model.

In the experiment, the DST working condition data at different temperatures are used as the training data of the BP neural network (BPNN), in which 70% are randomly selected as the training set, 15% as the verification set, and 15% as the test set. Set the number of training times (epochs) of the BP to 800, the number of hidden layer nodes to 8, the learning rate to 0.001, and the training target error to 1e-10. After the BPNN is successfully trained, the data of the HPPC working condition and the data of the BBDST working condition at the corresponding temperature are respectively employed for prediction and verification.

5.1. Co-estimation model validation of HPPC working condition

5.1.1. Validation under HPPC working condition at 15°C

The initial value of SOC is 1, and the initial value of SOH is 0.9524, which is the same as the reference value. The reference value of SOH can be obtained according to the capacity measurement method in Section 2.3.3, that is, after the battery is fully discharged three times, the average discharge capacity of the three times is calculated as the actual capacity of the lithium-ion battery in the current state, and the current capacity is obtained through the calculation formula of SOH Battery health status value. The experimental results are shown in Fig. 5-1.

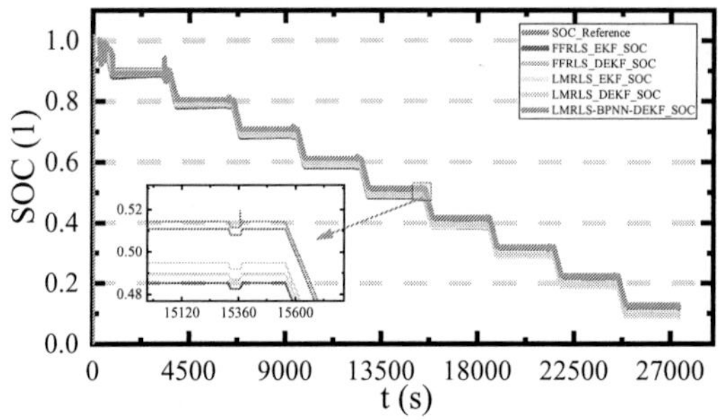

(a) Comparison of SOC estimation results of different algorithms

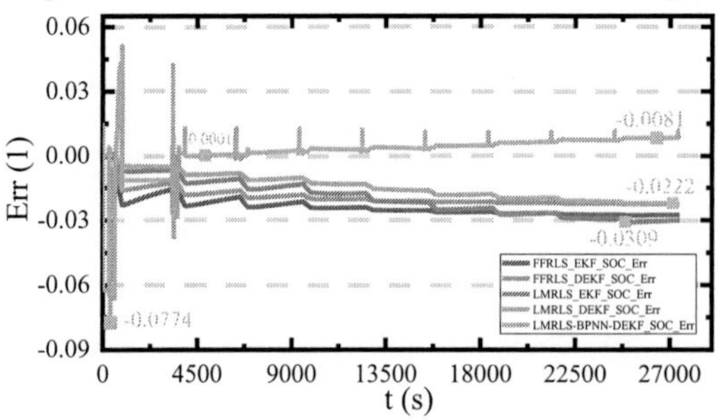

(b) Error comparison of SOC estimation results of different algorithms

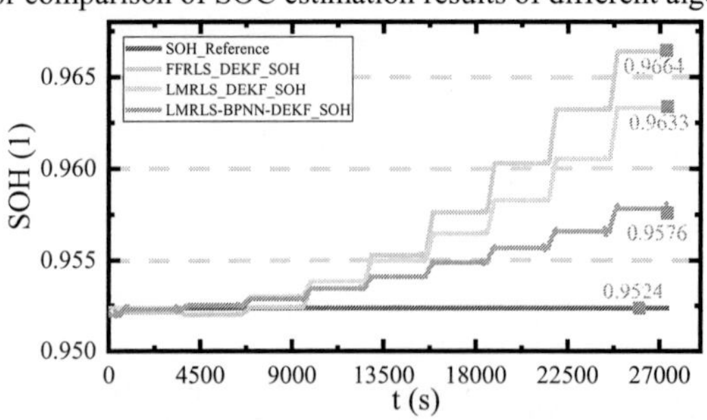

(c) Comparison of SOH estimation results of different algorithms

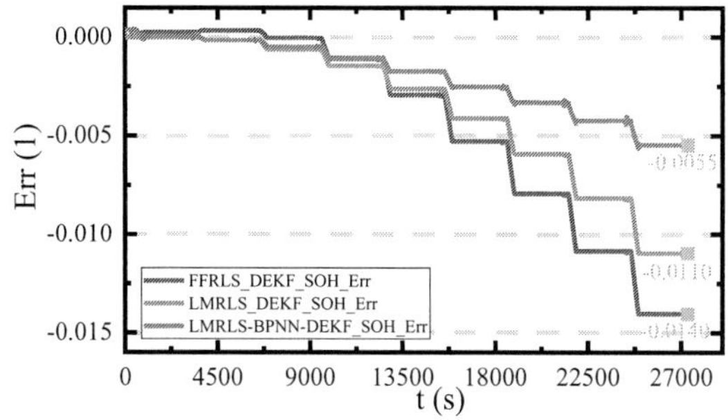

(d) Error comparison of SOH estimation results of different algorithms

Fig. 5-1 Validation results of SOC and SOH co-estimation from experimental data of HPPC working condition at 15°C

From Fig. 5-1 (a), it can be seen that the LMRLS-BPNN-DEKF algorithm has the best tracking effect in estimating SOC, and the advantage is more obvious compared with other algorithms, and then the LMRLS-DEKF, FFRLS-DEKF, LMRLS-EKF, and FFRLS-DEKF algorithms in order. It can be seen that the DEKF algorithm has a better co-estimation strategy, and the LMRLS algorithm also indirectly improves the estimation accuracy of SOC. As can be seen from Fig. 5-1 (b), the LMRLS-BPNN-DEKF algorithm has a large error fluctuation at first, with the maximum error being 7.74%, and then gradually converges to become smaller, and finally stabilizes around 0.81%. The DEKF algorithm is less error and more resistant to interference than the EKF. The LMRLS algorithm reduces the error of EKF and DEKF to a certain extent, and the overall effect is better than the FFRLS algorithm. However, the convergence and stability of the LMRLS-EKF algorithm are slightly worse than the FFRLS-EKF algorithm at the end of discharge. The comprehensive analysis of Fig. 5-1 (a) and Fig. 5-1 (b) shows that although the estimation error of the LMRLS-BPNN-DEKF algorithm is large in the beginning stage, its estimation effect gradually becomes better with the continuous learning and training of the BPNN, demonstrating the superiority of the model proposed in this study.

As can be seen from Fig. 5-1 (c) and Fig. 5-1 (d), the LMRLS algorithm and BP neural network algorithm have demonstrated significant improvement effects in the SOH estimation of lithium-ion batteries. Compared with the FFRLS-DEKF algorithm, the error of the LMRLS-DEKF algorithm is significantly reduced and the convergence is better, indicating that the LMRLS algorithm indirectly improves the accuracy of SOH estimation. Compared with other algorithms, the LMRLS-BPNN-DEKF algorithm has the smallest error and the highest stability, indicating that the BP neural network further improves the estimation accuracy and convergence effect of SOH.

Table 5-1 Error indicators of SOC and SOH co-estimation at 15℃ HPPC working condition

Estimation method	MAXE	MAE	RMSE
FFRLS-EKF(SOC)	2.77%	2.39%	2.43%
FFRLS-DEKF(SOC)	2.25%	1.95%	1.99%
LMRLS-EKF(SOC)	3.09%	1.98%	2.15%
LMRLS-DEKF(SOC)	2.22%	1.45%	1.57%
LMRLS-BPNN-DEKF(SOC)	7.74%	0.65%	1.11%
FFRLS-DEKF(SOH)	1.41%	0.45%	0.66%
LMRLS-DEKF(SOH)	1.10%	0.36%	0.51%
LMRLS-BPNN-DEKF(SOH)	0.57%	0.20%	0.27%

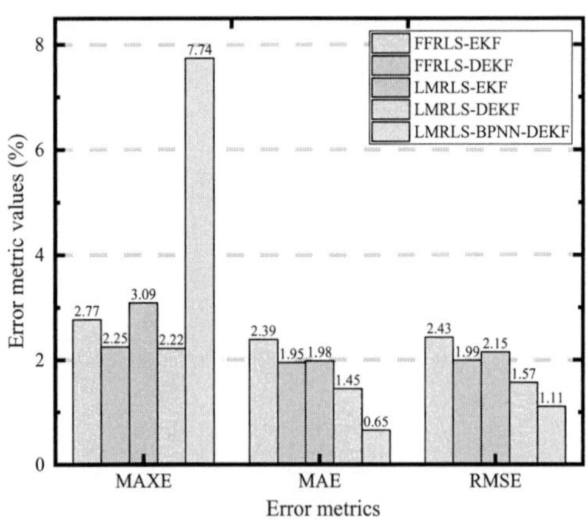

(a) Error indicators comparison of SOC estimation results

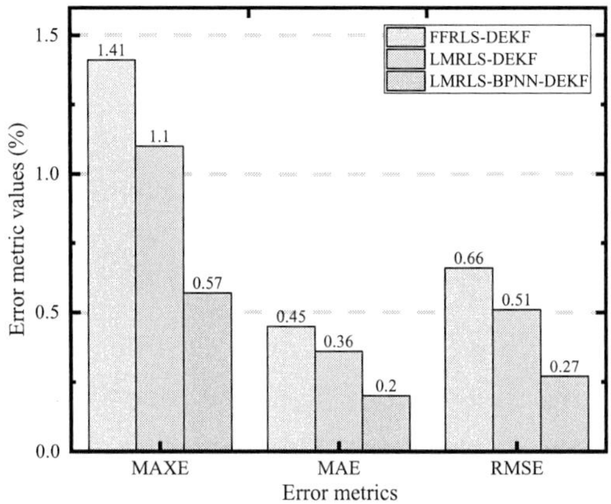

(b) Error indicators comparison of SOH estimation results

Fig. 5-2 Comparison chart of SOC and SOH co-estimation error indicators at 15°C HPPC working condition

From the overall comparison results in Table 5-1 and Fig. 5-2, it can be intuitively seen that the LMRLS-BPNN-DEKF algorithm proposed in this study shows a significant improvement in SOC and SOH estimation, with MAE and RMSE of SOC estimation error of 0.65% and 1.11%, respectively, which are 1.74% and 1.32% improvement compared to the FFRLS-EKF algorithm, respectively. It shows that the optimization effect of the BP neural network is more obvious. The LMRLS-EKF algorithm reduced the MAE and RMSE by 0.41% and 0.28%, respectively, compared with the FFRLS-EKF algorithm, and the LMRLS-DEKF algorithm reduced the MAE and RMSE by 0.53% and 0.58%, respectively, compared with the LMRLS-EKF algorithm. The above analysis shows that the method proposed in this study effectively improves the accuracy and robustness of SOC estimation. Moreover, the MAE and RMSE of the SOH estimation error of the proposed algorithm are 0.20% and 0.27%, respectively, with the maximum error reduced by 0.84% and 0.53%, MAE reduced by 0.25% and 0.16%, and RMSE reduced by 0.39% and 0.24%, respectively, compared with FFRLS-DEKF and LMRLS-DEKF, indicating that the method proposed in this study can effectively improve the accuracy of SOH estimation. In summary, the effectiveness of the LMRLS-BPNN-DEKF algorithm for improving accuracy and stability in the collaborative estimation of SOC and SOH at 15°C HPPC conditions has been verified.

5.1.2. Validation under HPPC working condition at 25°C

The initial value of SOC is 1 and the initial value of SOH is 0.9812, which is the same as the reference value. The experimental results are shown in Fig. 5-3.

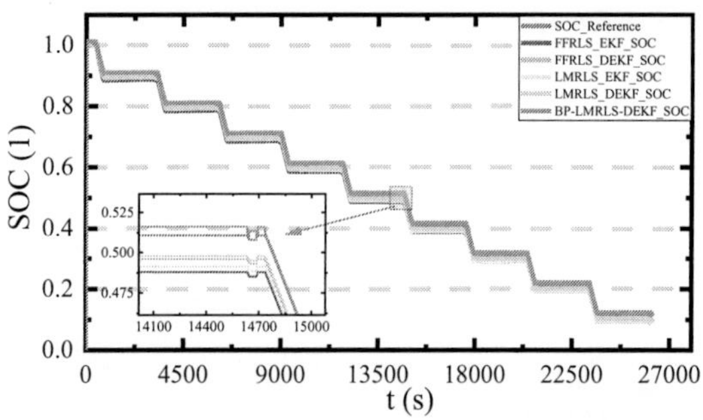

(a) Comparison of SOC estimation results of different algorithms

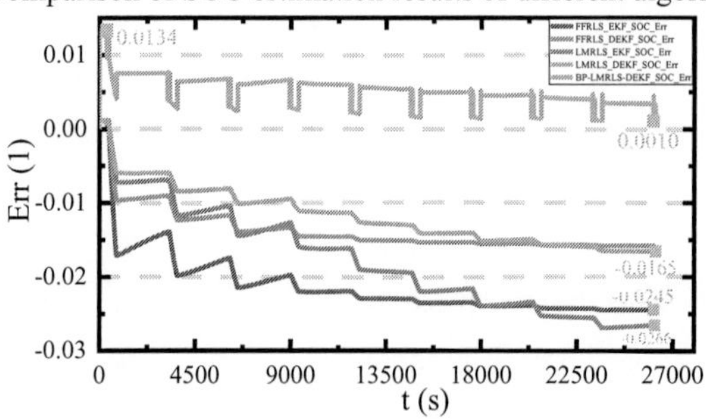

(b) Error comparison of SOC estimation results of different algorithms

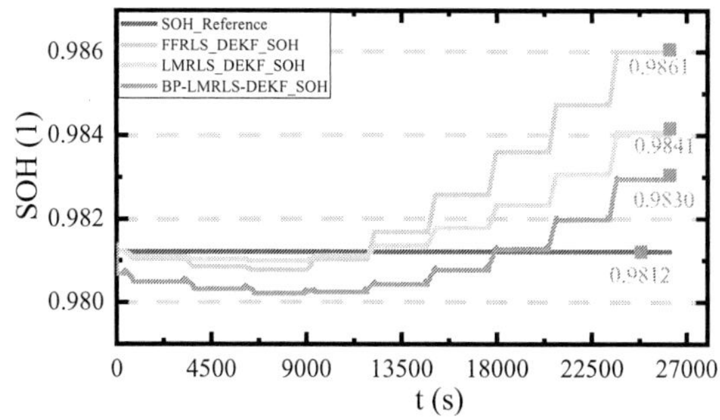

(c) Comparison of SOH estimation results of different algorithms

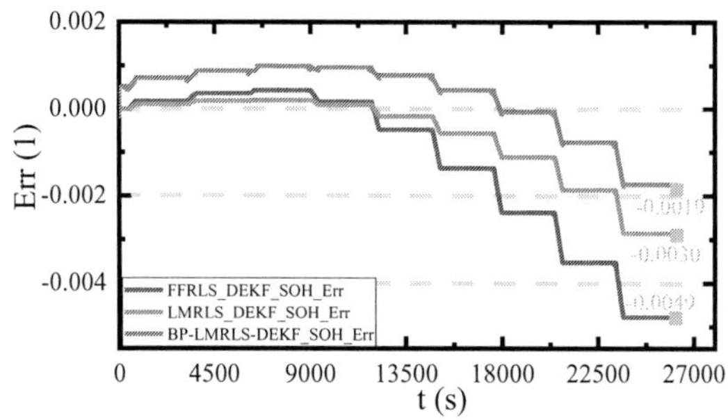

(d) Error comparison of SOH estimation results of different algorithms

Fig. 5-3 Validation results of SOC and SOH co-estimation from experimental data of HPPC working condition at 25°C

As can be seen from Fig. 5-3 (a), the SOC curve estimated by the LMRLS-BPNN-DEKF algorithm can follow the reference SOC curve better throughout the discharge process, especially at the discharge moment, with very high accuracy and stability. From Fig. 5-3 (b), it can be seen that the LMRLS-BPNN-DEKF algorithm has the best overall convergence, where the maximum error is 1.34% and the minimum error is 0.1%. Although the maximum error of LMRLS-DEKF is slightly larger than that of FFRLS-DEKF, it only exists at the end of discharge, and its overall error is significantly smaller than that of other algorithms except for the LMRLS-BPNN-DEKF algorithm. The comparison of the LMRLS-EKF algorithm and FFRLS-EKF algorithm is similar to the comparison of the LMRLS-

DEKF and FFRLS-DEKF, but it can be seen that the reduction of error is more obvious for the LMRLS algorithm compared with the FFRLS algorithm, and the improvement of convergence effect is more obvious for the DEKF algorithm compared with the EKF algorithm. In summary, the method model proposed in this study can effectively improve the estimation accuracy and convergence of SOC.

From Fig. 5-3 (c) and Fig. 5-3 (d), it can be seen that the LMRLS-BPNN-DEKF algorithm has the highest accuracy and stability in estimating SOH. Compared with the FFRLS-DEKF algorithm, the LMRLS-DEKF algorithm effectively improves the estimation accuracy of SOH, and the LMRLS-BPNN-DEKF algorithm further improves the accuracy and convergence of SOH on the basis of LMRLS-DEKF algorithm. From the above analysis, the LMRLS-BPNN-DEKF collaborative estimation model proposed in this study can effectively improve SOH estimation results.

Table 5-2 Error indicators of SOC and SOH co-estimation at 25°C HPPC working condition

Estimation method	**MAXE**	**MAE**	**RMSE**
FFRLS-EKF(SOC)	2.45%	2.13%	2.17%
FFRLS-DEKF(SOC)	1.58%	1.38%	1.41%
LMRLS-EKF(SOC)	2.69%	1.78%	1.91%
LMRLS-DEKF(SOC)	1.65%	1.19%	1.25%
LMRLS-BPNN-DEKF(SOC)	1.34%	0.52%	0.55%
FFRLS-DEKF(SOH)	0.49%	0.15%	0.21%
LMRLS-DEKF(SOH)	0.30%	0.07%	0.12%
LMRLS-BPNN-DEKF(SOH)	0.19%	0.07%	0.09%

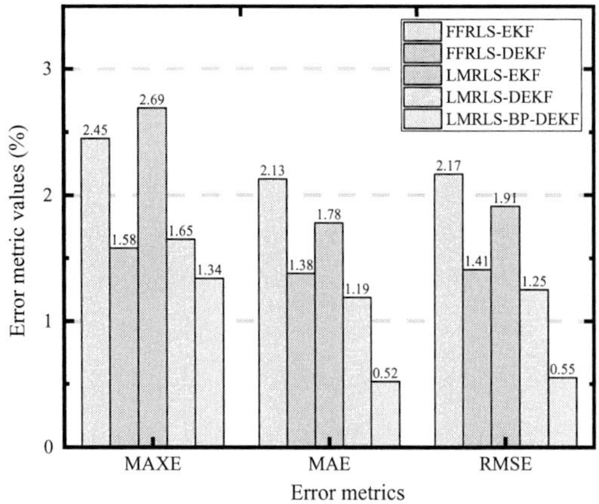

(a) Error indicators comparison of SOC estimation results

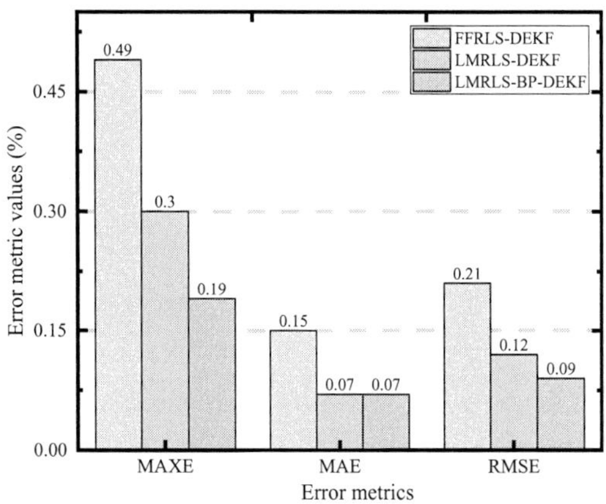

(b) Error indicators comparison of SOH estimation results

Fig. 5-4 Comparison chart of SOC and SOH co-estimation error indicators at 25°C HPPC working condition

From Table 5-2 and Fig. 5-4, it can be intuitively seen that the method model proposed in this study can effectively improve the estimation accuracy of the SOC and SOH of lithium batteries under HPPC operating conditions at 25°C. The MAXE, MAE, and RMSE of prediction SOC of the LMRLS-BPNN-DEKF algorithm are 1.34%, 0.52%, and 0.55% respectively, and the MAXE, MAE, and RMSE of prediction SOH is 0.19%, 0.07% and 0.09% respectively. It can be seen that the algorithm proposed in this study has high accuracy and robustness in both SOC and SOH prediction. Compared with the LMRLS-DEKF algorithm, the LMRLS-BPNN-DEKF algorithm reduces the MAXE of SOC estimation by 0.31%, the MAE by 0.67%, and the RMSE by 0.7%, and the MAXE of SOH estimation is reduced by 0.11% and the RMSE is reduced by 0.03%. It shows that the BP neural network introduced in this study can effectively improve the accuracy and robustness of SOC estimation and SOH estimation.

Comparing LMRLS-DEKF with LMRLS-EKF, it can be seen that the maximum error, MAE, and RMSE of SOC estimation by DEKF relative to EKF are reduced by 1.04%, 0.67%, and 0.7%, respectively. It can be seen that the co-estimation strategy proposed in this study can significantly improve SOC estimation accuracy and convergence. Comparing the LMRLS-EKF with the FFRLS-EKF, it can be seen that the maximum error of SOC estimated by the LMRLS-EKF is slightly larger, but the MAE and RMSE are reduced by 0.35% and 0.26%, respectively, and then combined with Fig. 5-4 (a) and (b), it can be seen that the estimation results of the LMELS-EKF have a certain degree of improvement compared with the FFRLS-EKF, indicating that the LMRLS parameter identification method proposed in this study can indirectly improve the estimation accuracy of SOC.

5.1.3. Validation under HPPC working condition at 35°C

The initial value of SOC is 1 and the initial value of SOH is 0.9873, which is the same as the reference value. The experimental results are shown in Fig. 5-5.

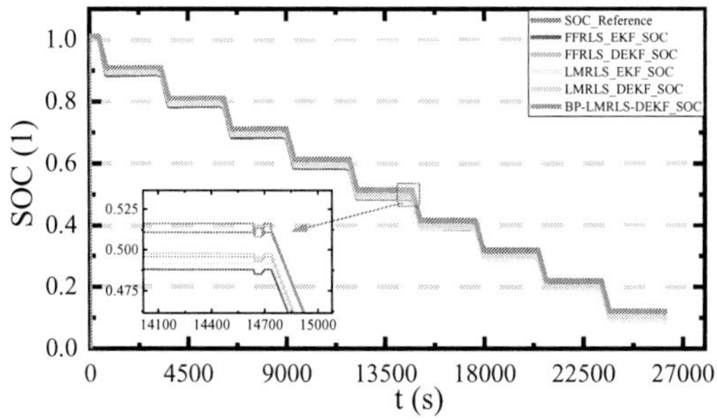

(a) Comparison of SOC estimation results of different algorithms

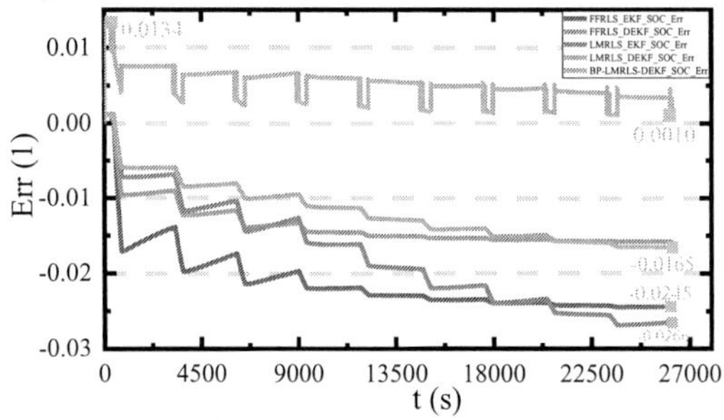

(b) Error comparison of SOC estimation results of different

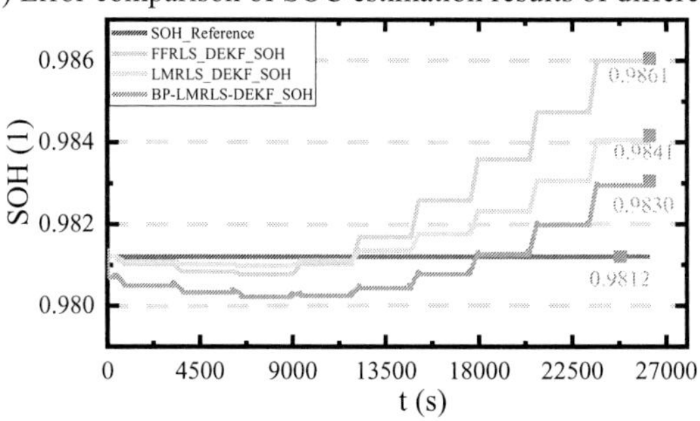

(c) Comparison of SOH estimation results of different algorithms

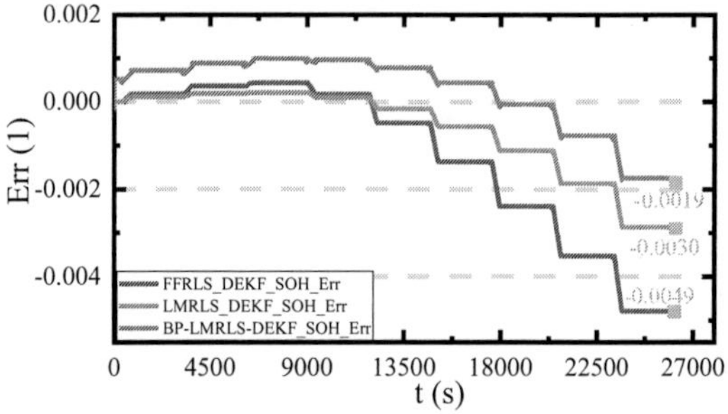

(d) Error comparison of SOH estimation results of different algorithms

Fig. 5-5 Validation results of SOC and SOH co-estimation from experimental data of HPPC working condition at 35°C

As can be seen from Fig. 5-5 (a), the LMRLS-BPNN-DEKF algorithm proposed in this study still has the best tracking effect in predicting the SOC curve at relatively high temperatures, which is the same as the above validation results at 15°C and 25°C, indicating that the proposed method in this study has better correction capability and stability at each temperature. It can be seen from Fig. 5-5 (b), the LMRLS-BPNN-DEKF algorithm has a large error in the initial stage but eventually stabilizes around 0, which is significantly smaller compared to the errors of other algorithms and also reflects its superiority.

Table 5-3 Error indicators of SOC and SOH co-estimation at 35°C HPPC working condition

Estimation method	**MAXE**	**MAE**	**RMSE**
FFRLS-EKF(SOC)	2.34%	2.03%	2.07%
FFRLS-DEKF(SOC)	1.51%	1.33%	1.36%
LMRLS-EKF(SOC)	2.50%	1.60%	1.73%
LMRLS-DEKF(SOC)	1.47%	0.99%	1.06%
LMRLS-BPNN-DEKF(SOC)	1.88%	0.13%	0.18%
FFRLS-DEKF(SOH)	0.30%	0.09%	0.12%
LMRLS-DEKF(SOH)	0.12%	0.03%	0.04%
LMRLS-BPNN-DEKF(SOH)	0.12%	0.02%	0.03%

From Fig. 5-5 (c) and (d), it can be seen that the LMRLS-BPNN-DEKF algorithm predicts the best curve convergence for SOH as well, and the LMRLS-DEKF algorithm is the second best, thus indicating that the

LMRLS parameter identification method proposed in this study and the introduced BP neural network can effectively improve the accuracy and convergence of SOH. In summary, the LMRLS-BPNN-DEKF synergistic prediction model proposed in this study is validated under 35°C HPPC conditions, and the results showed that the accuracy and stability of SOC and SOH prediction remained high at higher temperatures.

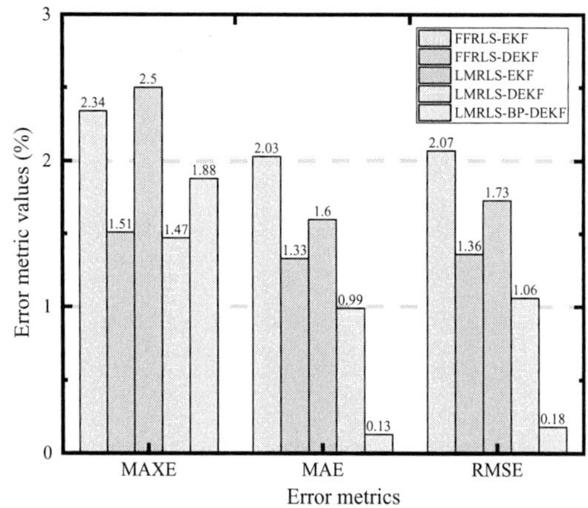

(a) Error indicators comparison of SOC estimation results

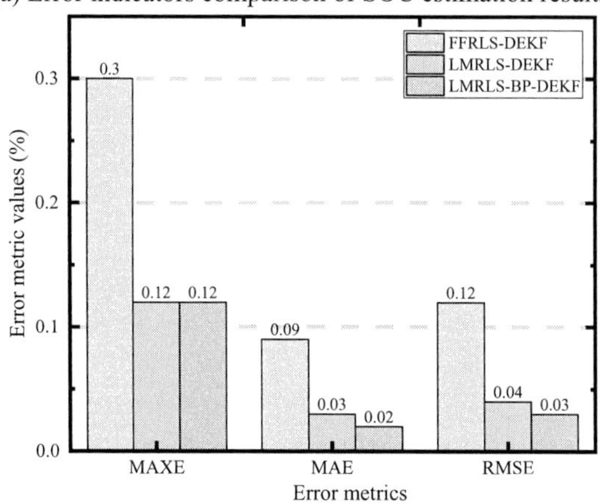

(b) Error indicators comparison of SOH estimation results

Fig. 5-6 Comparison chart of SOC and SOH co-estimation error indicators at 35°C HPPC working condition

From Table 5-3 and Fig. 5-6, it is visually clear that the LMRLS-BPNN-DEKF algorithm proposed in this study can still maintain high accuracy and stability for SOC and SOH estimation under HPPC working condition at 35°C. The MAE and RMSE of the LMRLS-BPNN-DEKF algorithm for estimating SOC are 0.13% and 0.18% respectively, which are 1.90% and 1.89% lower than those of the FFRLS algorithm, 1.47% and 1.55% lower than those of the LMRLS algorithm, and 0.86% and 0.88% lower than those of the LMRLS-DEKF algorithm. This shows that the method proposed in this study can effectively improve the estimation accuracy and convergence of SOC. Further analysis of the errors in estimating SOH by the LMRLS-BPNN-DEKF algorithm shows that the MAE and RMSE are 0.02% and 0.03% respectively, which are 0.07% and 0.09% lower compared to the FFRLS-DEKF algorithm and 0.01% lower compared to the LMRLS-DEKF algorithm. This indicates that the collaborative estimation model proposed in this study can improve the estimation accuracy of SOH to a certain extent. From the above analysis, it can be seen that the LMRLS parameter identification method designed, the proposed dual Kalman collaborative estimation strategy, and the BP neural network correction method introduced in this study can effectively improve the accuracy and robustness of SOC and SOH estimation even at higher temperatures.

5.2. Co-estimation model validation of BBDST working condition

To further verify the effectiveness of the proposed algorithm for improving the synergistic estimation of SOC and SOH in this study, the algorithm is validated under more complex BBDST working conditions to analyze the adaptive adjustment capability of the proposed model approach for more complex conditions.

5.2.1. Validation under BBDST working condition at 15°C

The initial value of SOC is 1 and the initial value of SOH is 0.9524, which is the same as the reference value. The experimental results are shown in Fig. 5-7.

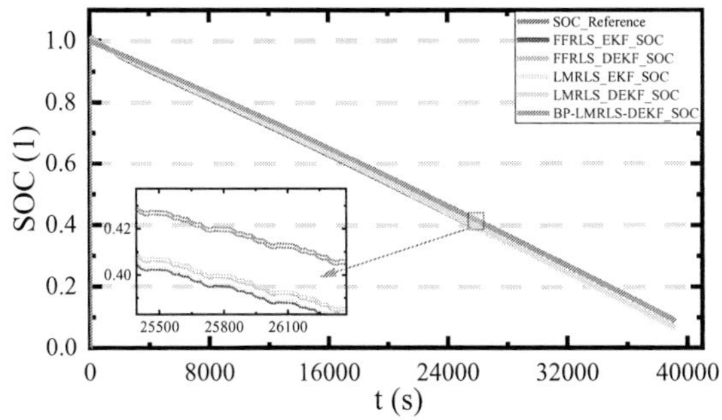

(a) Comparison of SOC estimation results of different algorithms

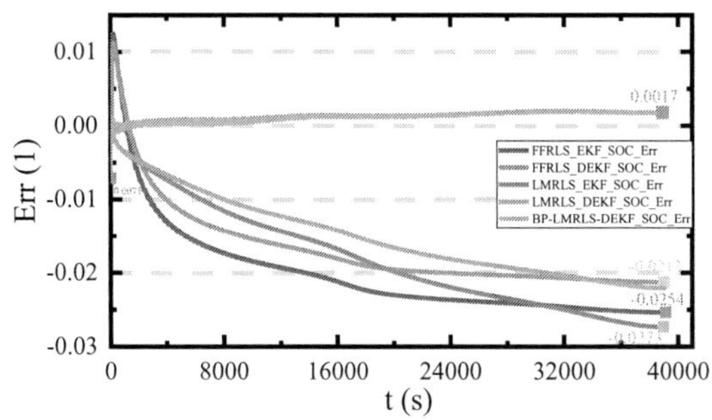

(b) Error comparison of SOC estimation results of different algorithms

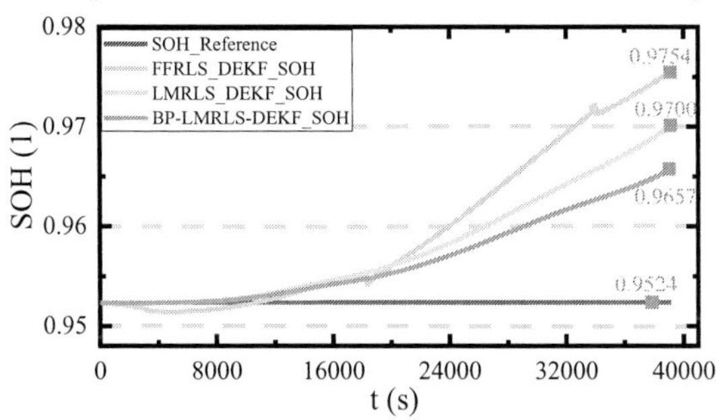

(c) Comparison of SOH estimation results of different algorithms

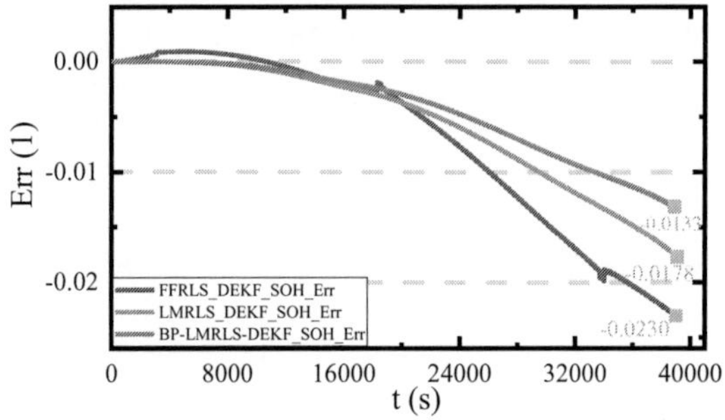

(d) Error comparison of SOH estimation results of different algorithms

Fig. 5-7 Validation results of SOC and SOH co-estimation from experimental data of BBDST working condition at 15°C

From Fig. 5-7 (a), it can be seen that the LMRLS-BPNN-DEKF algorithm has better tracking performance and its tracking ability is significantly better than the other algorithms. Among them, the FFRLS-EKF tracking is the worst, and the LMRLS-DEKF is the second best. As can be seen from Fig. 5-7 (b), the maximum error of the LMRLS-BPNN-DEKF algorithm is only 0.71%, and the error fluctuates greatly only in the initial stage. With the learning and training of the BP neural network, the training error soon converges to finally stabilize around 0.17%. The overall error of LMRLS-DEKF and LMRLS-EKF is significantly smaller than that of FFRLS-DEKF and FFRLS-EKF, and the error is only slightly larger at the end of discharge. This shows that the LMRLS parameter identification method proposed in this study can indirectly improve the estimation accuracy of SOC. Compared with the EKF algorithm, the DEKF algorithm has a faster convergence speed and smaller error, indicating that the collaborative estimation strategy proposed in this study can effectively improve the accuracy and robustness of SOC estimation.

From Fig. 5-7 (c) and Fig. 5-7 (d), it can be seen that the LMRLS-BPNN-DEKF collaborative estimation algorithm has certainly improved the estimation results of SOH for lithium batteries. Compared with the FFRLS-DEKF and LMRLS-DEKF algorithms, the LMRLS-BPNN-DEKF algorithm has a faster convergence speed and a better convergence effect, indicating that the LMRLS algorithm and BP neural network can effectively improve the accuracy of lithium battery SOH estimation results and improve its convergence effect.

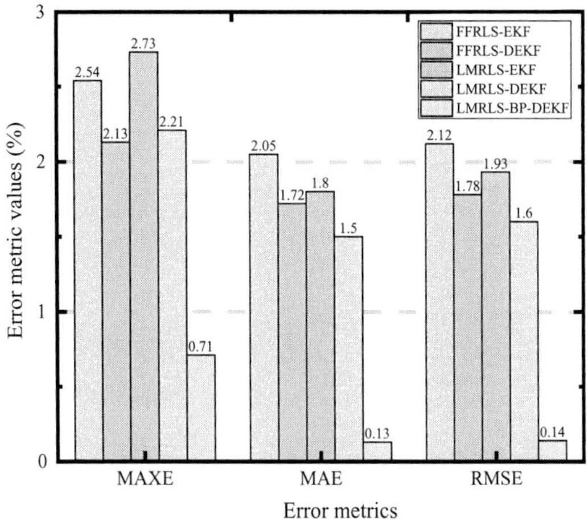

(a) Error indicators comparison of SOC estimation results

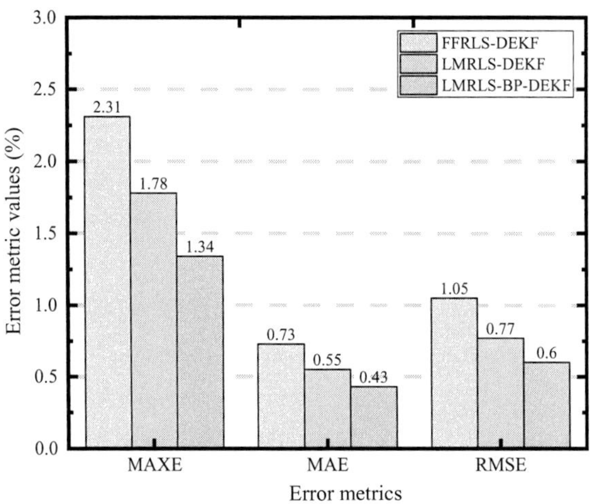

(b) Error indicators comparison of SOH estimation results

Fig. 5-8 Comparison chart of SOC and SOH co-estimation error indicators at 15°C BBDST working condition

Table 5-4 Error indicators of SOC and SOH co-estimation at 15°C BBDST working condition

Estimation method	**MAXE**	**MAE**	**RMSE**
FFRLS-EKF(SOC)	2.54%	2.05%	2.12%
FFRLS-DEKF(SOC)	2.13%	1.72%	1.78%
LMRLS-EKF(SOC)	2.73%	1.80%	1.93%
LMRLS-DEKF(SOC)	2.21%	1.50%	1.60%
LMRLS-BPNN-DEKF(SOC)	0.71%	0.13%	0.14%
FFRLS-DEKF(SOH)	2.31%	0.73%	1.05%
LMRLS-DEKF(SOH)	1.78%	0.55%	0.77%
LMRLS-BPNN-DEKF(SOH)	1.34%	0.43%	0.60%

It can be seen intuitively from Table 5-4 and Fig. 5-8 that the LMRLS-BPNN-DEKF collaborative estimation model proposed in this study has shown good improvement effects in the estimation of lithium battery SOC and SOH. In SOC estimation, the maximum error of the LMRLS-BPNN-DEKF algorithm is only 0.71%, and the MAE and RMSE are 0.13% and 0.14% respectively, which is a very obvious improvement compared with other algorithms and show very good high accuracy and high stability. Among them, the MAXE, MAE, and RMSE of LMRLS-BPNN-DEKF are reduced by 1.83%, 1.92%, and 1.98% respectively compared to FFRLS-EKF, indicating that the LMRLS algorithm and BP neural network algorithm have remarkable optimization effect. The MAXE, MAE, and RMSE of FFRLS-DEKF are reduced by 0.41%, 0.33%, and 0.34%, respectively, compared to FFRLS-EKF, indicating that the co-estimation strategy can improve the estimation results of SOC. The MAXE, MAE, and RMSE of LMRLS-DEKF are reduced by 0.08%, 0.22%, and 0.52% respectively compared to FFRLS-DEKF, indicating that the LMRLS algorithm can effectively improve the accuracy of parameter identification and hence the accuracy and stability of SOC estimation.

In the estimation of SOH, the MAXE, MAE, and RMSE of the LMRLS-BPNN-DEKF algorithm are reduced by 0.97%, 0.3%, and 0.45%, respectively, compared to the FFRLS-DEKF algorithm, and by 0.44%, 0.12 and 0.17%, respectively, compared to the LMRLS-DEKF, thus indicating that the collaborative estimation strategy proposed in this study can effectively improve the accuracy and robustness of SOH. The above detailed analysis of the estimation results of SOC and SOH of lithium battery by LMRLS-BPNN-DEKF algorithm under BBDST working conditions at 15°C, which shows that the method proposed in this study has high accuracy and robustness.

5.2.2. Validation under BBDST working condition at 25°C

The initial SOC value is 1 and the initial SOH value is 0.9812, which is the same as the reference value. The experimental results are shown in Fig. 5-9.

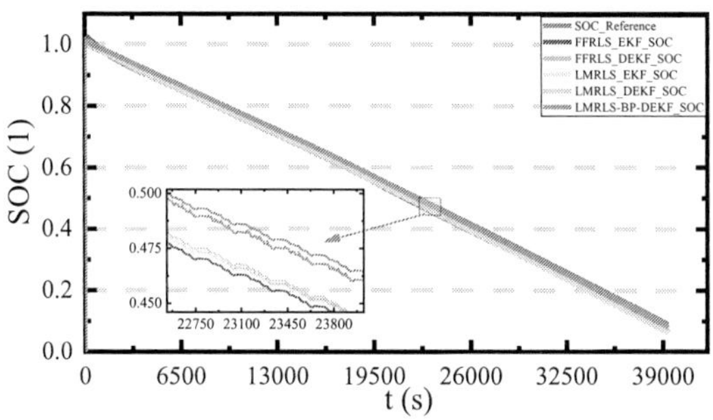

(a) Comparison of SOC estimation results of different algorithms

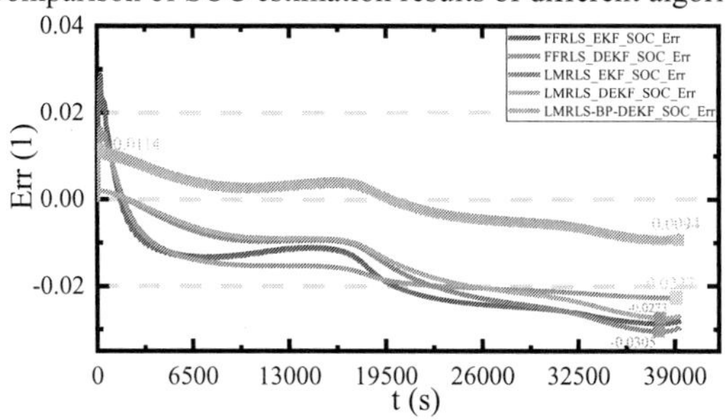

(b) Error comparison of SOC estimation results of different algorithms

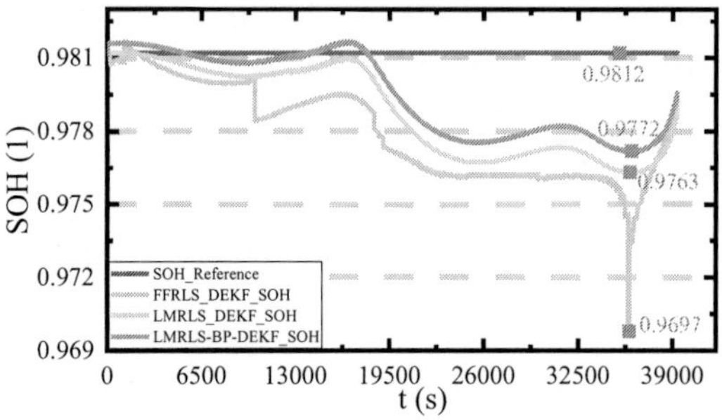

(c) Comparison of SOH estimation results of different algorithms

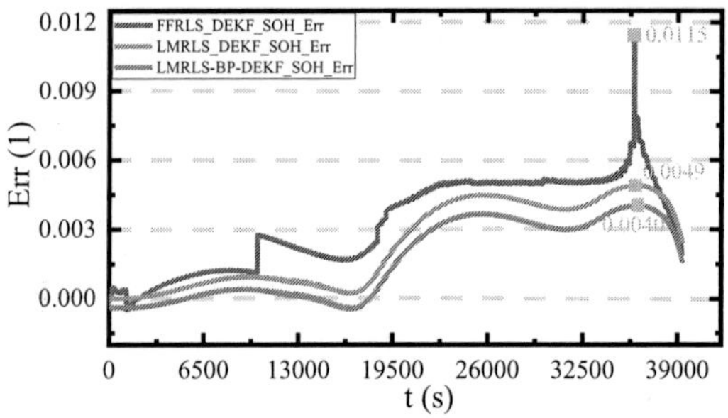

(d) Error comparison of SOH estimation results of different algorithms

Fig. 5-9 Validation results of SOC and SOH co-estimation from experimental data of BBDST working condition at 25°C

From Fig. 5-9 (a), it can be seen that the LMRLS-BPNN-DEKF algorithm proposed in this study has higher accuracy and stability in estimating SOC during the whole discharge process. From Fig. 5-9 (b), we can see that the maximum error of LMRLS-BPNN-DEKF is 1.14%, the maximum error of FFRLS-DEKF is 2.27%, the maximum error of FFRLS-EKF and LMRLS-DEKF is around 2.75%, and the maximum error of LMRLS-EKF is 3.05%. On the whole, the LMRLS algorithm has the best convergence and small error; the overall error of the LMRLS-DEKF and LMRLS-EKF algorithms is small, and the error is only slightly larger at the end of discharge; the LMRLS-DEKF algorithm has better overall error and convergence than the LMRLS-EKF algorithm, especially the late error is significantly smaller. In summary, the LMRLS parameter identification method, the dual Kalman co-estimation strategy, and the BP neural network algorithm proposed in this study can all effectively improve the estimation accuracy of SOC and optimize its convergence effect.

As can be seen in Fig. 5-9 (c) and Fig. 5-9 (d), the LMRLS-BPNN-DEKF algorithm also provides significant improvements in the estimation of SOH. The FFRLS-DEKF algorithm predicts SOH with large curve fluctuations and poor convergence, while LMRLS-DEKF and LMRLS-BPNN-DEKF have better convergence and relatively small errors. Comparing the LMRLS-BPNN-DEKF algorithm with the LMRLS-DEKF algorithm, it can be seen that the error of the LMRLS-BPNN-DEKF algorithm is further reduced and the tracking effect of the SOH prediction curve is further improved. This shows that the method proposed in this study can effectively improve the accuracy and stability of SOH estimation.

Table 5-5 Error indicators of SOC and SOH co-estimation at 25°C BBDST working condition

Estimation method	**MAXE**	**MAE**	**RMSE**
FFRLS-EKF(SOC)	2.87%	1.87%	2.01%
FFRLS-DEKF(SOC)	2.27%	1.73%	1.80%
LMRLS-EKF(SOC)	3.05%	1.61%	1.86%
LMRLS-DEKF(SOC)	2.75%	1.45%	1.68%
LMRLS-BPNN-DEKF(SOC)	1.14%	0.53%	0.59%
FFRLS-DEKF(SOH)	1.15%	0.33%	0.38%
LMRLS-DEKF(SOH)	0.49%	0.24%	0.30%
LMRLS-BPNN-DEKF(SOH)	0.40%	0.18%	0.24%

From the comparison of the error metrics of different algorithms in Table 5-5 and Fig. 5-10, it can be seen that the LMRLS-BPNN-DEKF collaborative estimation model proposed in this study has a high accuracy in predicting both SOC and SOH of lithium batteries under the BBDST working condition at 25°C. The MAXE, MAE, and RMSE of SOC estimated by the LMRLS-BPNN-DEKF algorithm are 1.14%, 0.53%, and 0.59%, respectively. The MAXE, MAE, and RMSE of SOH prediction are 0.40%, 0.18%, and 0.24%, respectively.

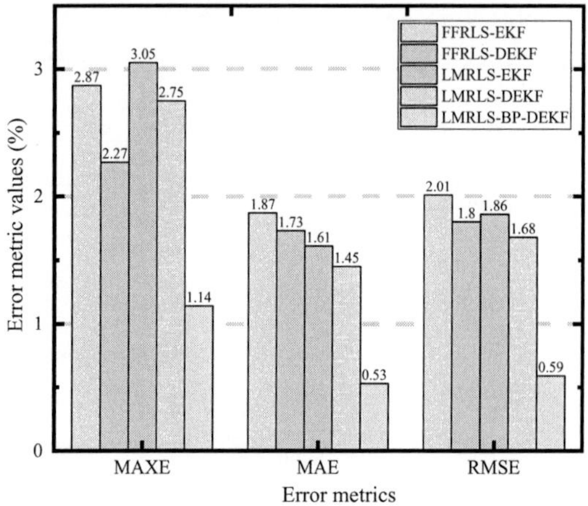

(a) Error indicators comparison of SOC estimation results

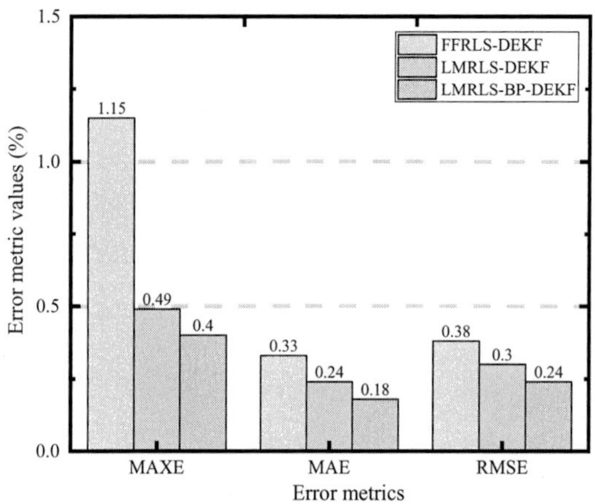

(b) Error indicators comparison of SOH estimation results

Fig. 5-10 Comparison chart of SOC and SOH co-estimation error indicators at 25°C BBDST working condition

Relative to the conventional FFRLS-EKF algorithm, the MAXE, MAE, and RMSE of SOC prediction are reduced by 1.73%, 1.34%, and 1.42% respectively. Relative to the FFRLS-DEKF algorithm, the MAXE, MAE, and RMSE of SOC estimation are reduced by 1.13%, 1.20%, and 1.21%, respectively, and the MAXE, MAE, and RMSE of SOH prediction are reduced by 0.75%, 0.15%, and 0.14%, respectively. This shows that the LMRLS-BPNN-DEKF algorithm proposed in this study can effectively improve the prediction accuracy and tracking performance of SOC and SOH.

Further from the comparison of the LMRLS-DEKF algorithm with the FFRLS-DEKF algorithm, it can be seen that the MAE and RMSE of the LMRLS-DEKF algorithm for predicting SOC are reduced by 0.28% and 0.12%, respectively, compared with the FFRLS-DEKF algorithm, and the MAE and RMSE of predicting SOH are reduced by 0.09% and 0.08%, respectively. This shows that the LMRLS parameter identification method proposed in this study can improve the prediction accuracy and stability of SOC and SOH to a certain extent. According to the comparison between the LMRLS-BPNN-DEKF algorithm and the LMRLS-DEKF algorithm, the MAE and RMSE of the LMRLS-BPNN-DEKF algorithm decreased by 0.92% and 1.09% respectively in predicting SOC, and the MAE and RMSE of SOH decreased by 0.06% compared to the LMRLS-DEKF algorithm. This indicates that the BP neural network algorithm introduced in this study can effectively improve the estimation accuracy of SOC and improve the estimation results of SOH to a certain extent.

5.2.3. Validation under BBDST working condition at 35°C

The initial value of SOC is 1, and the initial value of SOH is 0.9873, which is the same as the reference value. The experimental results are shown in Fig. 5-11.

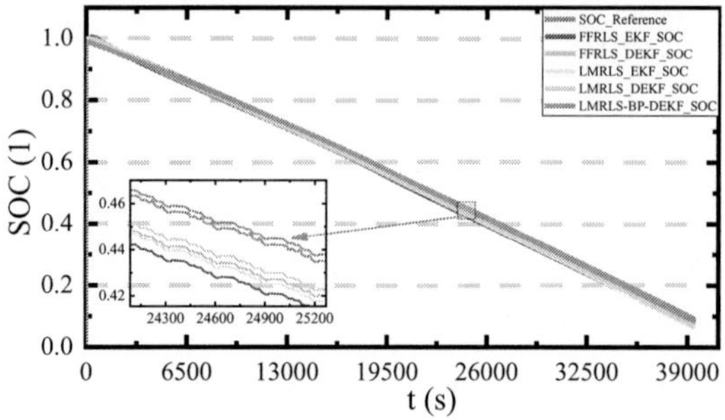

(a) Comparison of SOC estimation results of different algorithms

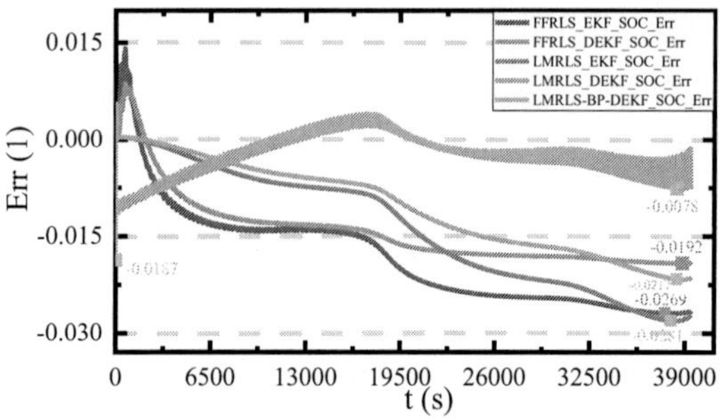

(b) Error comparison of SOC estimation results of different algorithms

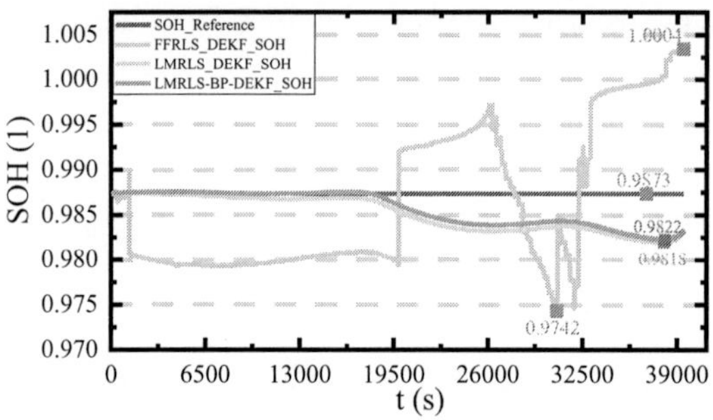

(c) Comparison of SOH estimation results of different algorithms

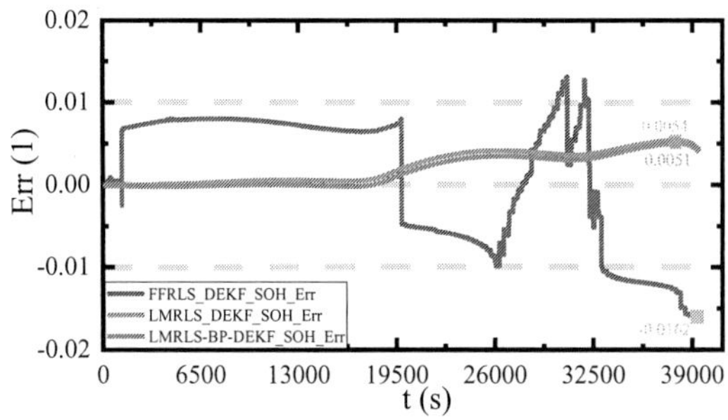

(d) Error comparison of SOH estimation results of different algorithms

Fig. 5-11 Validation results of SOC and SOH co-estimation from experimental data of BBDST working condition at 35°C

From Fig. 5-11 (a), it can be seen that the curve tracking effect of the LMRLS-BPNN-DEKF algorithm proposed in this study is still optimal in predicting SOC even under the BBDST operating condition with high temperatures. As can be seen from Fig. 5-11 (b), the error of the LMRLS-BPNN-DEKF algorithm is small throughout the discharge phase, where the maximum error is 1.87% and finally stabilizes at about 0.78%. On the whole, compared to the FFRLS-EKF and LMRLS-EKF algorithms, the FFRLS-DEKF and LMRLS-DEKF algorithms converge better and have relatively smaller errors. Comparing the FFRLS-EKF and LMRLS-EKF algorithms, it can be seen that the overall error of LMRLS-EKF is relatively smaller, with only slightly larger in the final stage. From the above analysis, it can be seen that this is similar to the verification results of the operating conditions at 15°C and 25°C, which further indicates that the method and strategy proposed in this study can effectively improve the prediction results of SOC.

From Fig. 5-11 (c) and Fig. 5-11 (d), it can be seen that the LMRLS-BPNN-DEKF algorithm predicts the curve of SOH with the best following effect in the first half of the stage, and the accuracy in the second half stage is also higher. On the whole, the convergence effect and precision of the LMRLS-BPNN-DEKF algorithm are quite ideal. The LMRLS-DEKF algorithm is next, and the accuracy and convergence of the FFRLS-DEKF algorithm are relatively poor. It shows that the algorithm model proposed in this study can still effectively improve SOH estimation results when the temperature is too high.

Table 5-6 Error indicators of SOC and SOH co-estimation at 35°C BBDST working condition

Estimation method	MAXE	MAE	RMSE
FFRLS-EKF(SOC)	2.69%	1.87%	1.98%
FFRLS-DEKF(SOC)	1.92%	1.46%	1.53%
LMRLS-EKF(SOC)	2.81%	1.37%	1.64%
LMRLS-DEKF(SOC)	2.17%	1.05%	1.25%
LMRLS-BPNN-DEKF(SOC)	1.87%	0.36%	0.43%
FFRLS-DEKF(SOH)	1.62%	0.75%	0.81%
LMRLS-DEKF(SOH)	0.54%	0.22%	0.29%
LMRLS-BPNN-DEKF(SOH)	0.51%	0.18%	0.25%

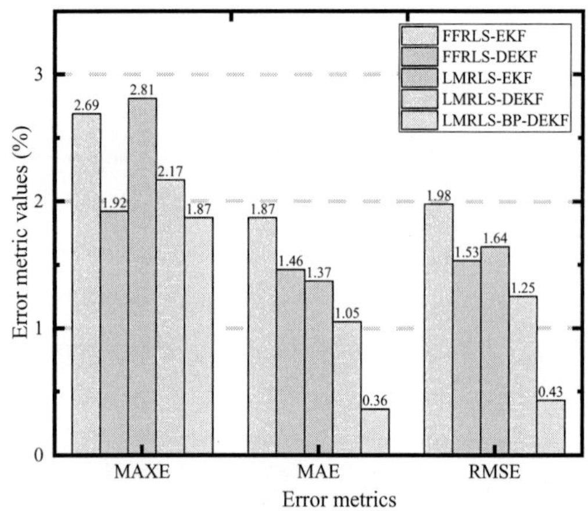

(a) Error indicators comparison of SOC estimation results

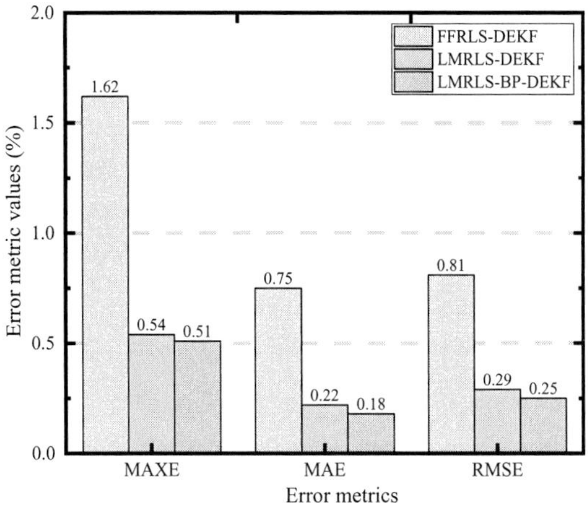

(b) Error indicators comparison of SOH estimation results

Fig. 5-12 Comparison chart of SOC and SOH co-estimation error indicators at 35°C BBDST working condition

From the further analysis of the error metrics of different algorithms in Table 5-6 and Fig. 5-12, it can be seen that the LMRLS-BPNN-DEKF collaborative estimation algorithm model proposed in this study can still maintain high accuracy and stability for SOC and SOH estimation results of the lithium-ion battery under BBDST operating conditions at 35°C. The MAE and RMSE of the LMRLS-BPNN-DEKF algorithm for predicting SOC error are 0.36% and 0.43%, and the MAE and RMSE for predicting SOH error are 0.18% and 0.25%, respectively. Compared with the LMRLS-DEKF algorithm, the MAE and RMSE of the LMRLS-BPNN-DEKF algorithm are reduced by 0.69% and 0.82% for SOC estimation and 0.04% for SOH estimation, thus indicating that the BP neural network can effectively improve the accuracy of SOC and SOH estimation.

Comparing the LMRLS-DEKF algorithm with the LMRLS-EKF algorithm, it can be seen that the MAE and RMSE of the predicted SOC of the LMRLS-DEKF algorithm are reduced by 0.32% and 0.39%, respectively, which shows that the dual Kalman synergistic estimation strategy proposed in this study can effectively improve the estimation results of SOC. Comparing the LMRLS-EKF algorithm with the FFRLS-EKF algorithm, it can be seen that the MAE and RMSE of the LMRLS-EKF predicted SOC are reduced by 0.50% and 0.34%, respectively, which shows that the LMRLS algorithm indirectly improves the accuracy of SOC estimation. In summary, the designed LMRLS parameter identification algorithm, the proposed collaborative estimation strategy, and the introduced BP neural network algorithm in this study can effectively improve SOC estimation accuracy and improve SOH prediction accuracy and robustness to a certain extent even under complex operating conditions with high temperatures.

5.3. Validation analysis of different state initial values

To further analyze the robustness and correction capability of the LMRLS-BPNN-DEKF co-estimation model, SOC and SOH with different initial values are set for experimental validation. In this section, the HPPC, DST, and BBDST working condition data at 25 °C are used for verification analysis. The initial value of SOC is 0.8, and the initial value of SOH is 0.9757. The data under DST working conditions are used as the training data of the BP neural network, and 70% are randomly selected as the training set, 15% as the validation set, and 15% as the test set. The training number of the BP neural network is set to 800, the number of nodes in the hidden layer is set to 8, the learning rate is set to 0.001, and the training target error is set to 1e-10. After the BP neural network training achieves the target effect, the data of HPPC and BBDST working conditions are applied to verify the analysis.

5.3.1. Different initial values validation under HPPC working condition

The initial value of SOC at 25°C HPPC working condition is 0.8, the initial value of SOH is 0.9757 and the reference SOH value is 0.9812, the experimental results are shown in Fig. 5-13. Since the initial errors are all large, the maximum error (MAXE) is replaced with the mean error (ME) for easy comparison. The calculation formula of ME is shown in Eq. (5-1)

$$ME = \frac{1}{n}\sum_{i=1}^{n}(x_k - \hat{x}_k) \tag{5-1}$$

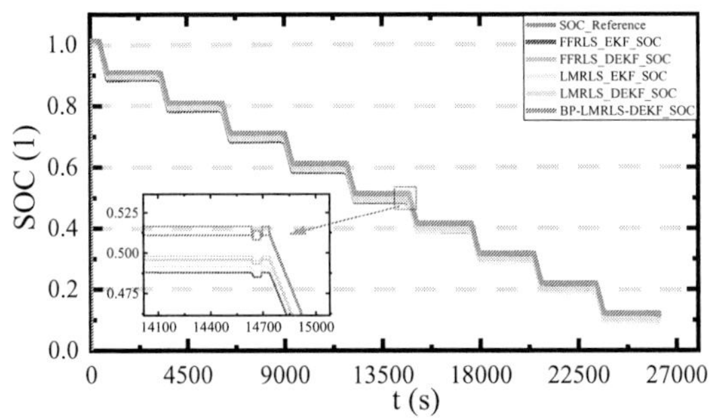

(a) Comparison of SOC estimation results of different algorithms

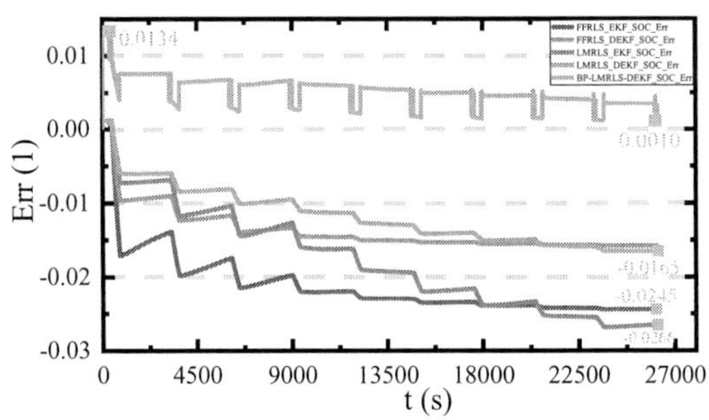

(b) Error comparison of SOC estimation results of different algorithms

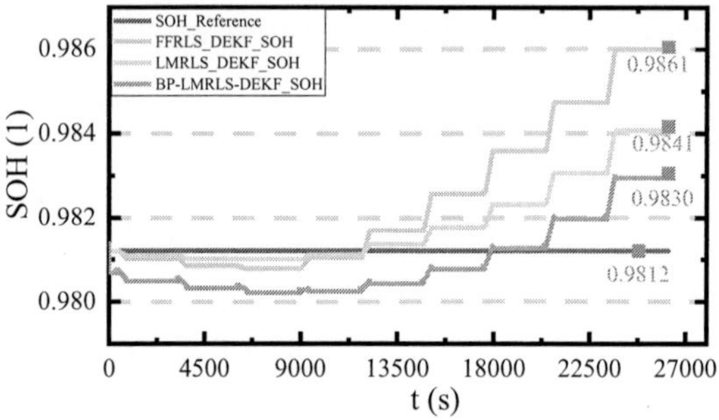

(c) Comparison of SOH estimation results of different algorithms

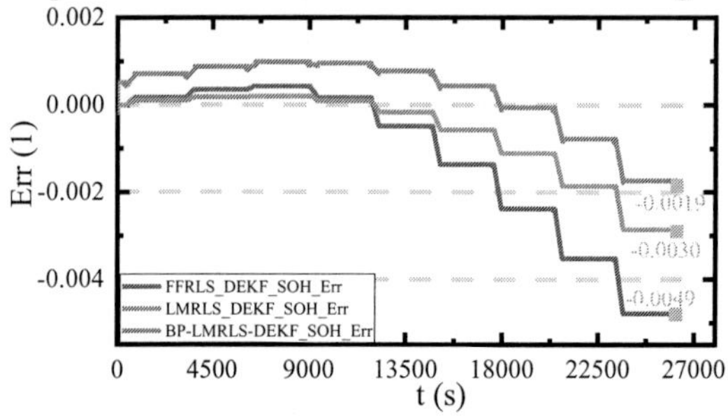

(d) Error comparison of SOH estimation results of different algorithms

Fig. 5-13 Verification results of SOC and SOH collaborative estimation of different state initial values under HPPC working conditions

As can be seen from Fig. 5-13, the LMRLS-BPNN-DEKF algorithm proposed in this study can still accurately predict SOC and SOH even if the initial values are set with large deviations. As can be seen from Fig. 5-13 (a) that when the initial value of the SOC setting deviates greatly from the real value, the LMRLS-BPNN-DEKF algorithm can quickly correct to the correct estimated value and maintain relatively high accuracy. From Fig. 5-13 (b), it can be seen that the LMRLS-BPNN-DEKF algorithm has a large error only in the beginning stage, and as the BP neural network keeps learning, the error becomes smaller and smaller and finally converges near the value of 0. From Fig. 5-13 (c) and Fig. 5-13 (d), it can be seen that the LMRLS-BPNN-DEKF algorithm can still predict the SOH values accurately when the initial SOH values deviate significantly. It can be seen from Fig. 5-13 (c) that the SOH prediction curve of the LMRLS-BPNN-DEKF algorithm has a good tracking effect during the entire discharge process, and it always follows near the reference value. As shown in Fig. 5-13 (d), the error of the LMRLS-BPNN-DEKF algorithm is relatively small. This shows that the method model proposed in this study still has high accuracy and stability when the initial value of the state is set with a large deviation.

Table 5-7 Error indicators of SOC and SOH co-estimation for different state initial values under HPPC working conditions

Estimation method	**MAXE**	**MAE**	**RMSE**
FFRLS-EKF(SOC)	2.04%	2.18%	2.29%
FFRLS-DEKF(SOC)	1.63%	1.67%	1.75%
LMRLS-EKF(SOC)	1.74%	1.77%	1.85%
LMRLS-DEKF(SOC)	1.13%	1.20%	1.27%
LMRLS-BPNN-DEKF(SOC)	0.19%	0.71%	0.98%
FFRLS-DEKF(SOH)	0.28%	0.55%	0.33%
LMRLS-DEKF(SOH)	0.25%	0.32%	0.36%
LMRLS-BPNN-DEKF(SOH)	0.11%	0.27%	0.31%

It can be further visualized from Table 5-7 and Fig. 5-14 that the LMRLS-BPNN-DEKF collaborative prediction algorithm proposed in this study still has high accuracy in the estimated values of SOC and SOH under different state initial values. Where, the mean error (ME) of the LMRLS-BPNN-DEKF algorithm in predicting SOC is 0.19%, MAE is 0.71% and RMSE is 0.98%. The ME of predicted SOH is 0.11%, MAE is 0.27%, and RMSE is 0.31%.

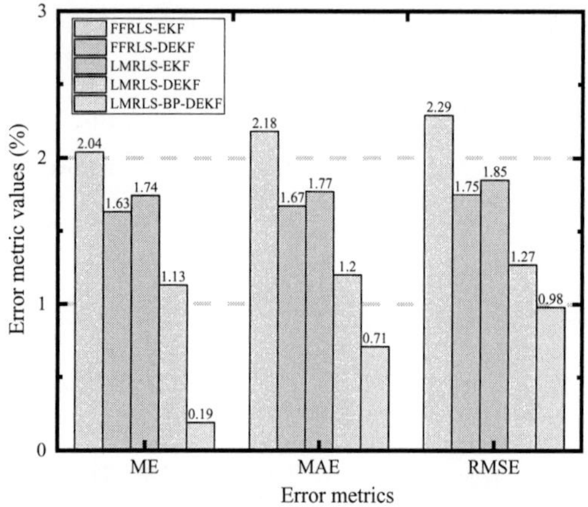

(a) Error indicators comparison of SOC estimation results

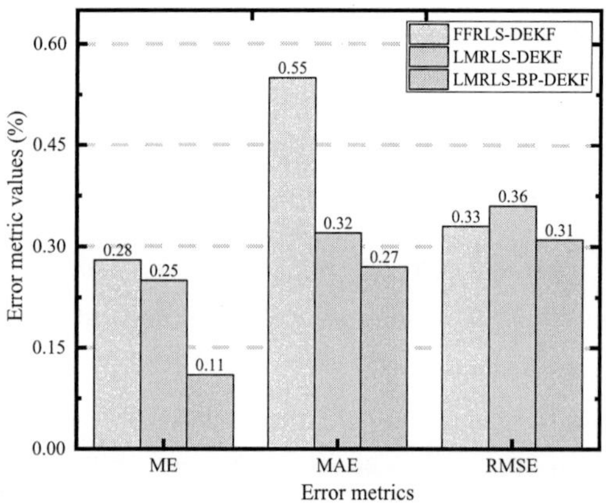

(b) Error indicators comparison of SOH estimation results

Fig. 5-14 Comparison chart of SOC and SOH co-estimation error indicators for different state initial values under HPPC working condition

Compared with other algorithms, it can be seen that the error indicators of the LMRLS-BPNN-DEKF algorithm are smaller, especially the improvement and optimization of the error metrics of SOC are more obvious. Compared with the FFRLS-EKF algorithm, the ME is reduced by 1.85%, and the MAE is reduced. 1.47%, RMSE decreased by 1.31%. In summary, the algorithm model proposed in this study has good calibration capability even when the initial value of the state deviates significantly.

5.3.2. Different initial values validation under BBDST working condition

Under BBDST working condition at 25°C, the initial SOC value is 0.8, the initial SOH value is 0.9757, and the reference SOH value is 0.9812. The experimental results are shown in Fig. 5-15. For ease of representation, ME is taken in absolute terms in Fig. 5-16.

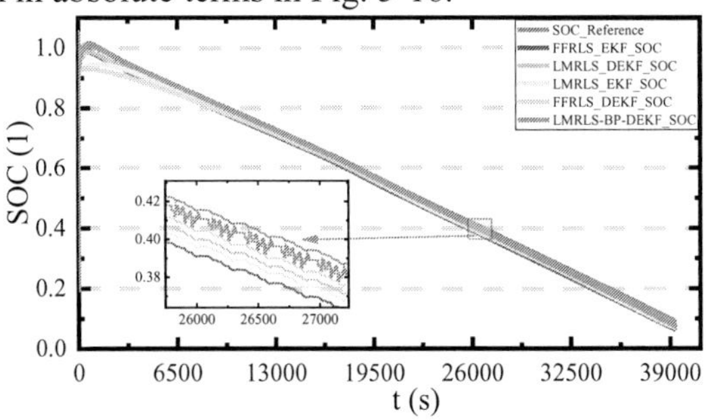

(a) Comparison of SOC estimation results of different algorithms

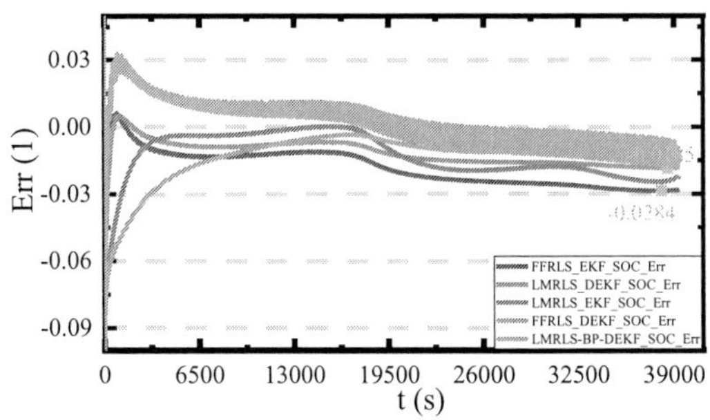

(b) Error comparison of SOC estimation results of different algorithms

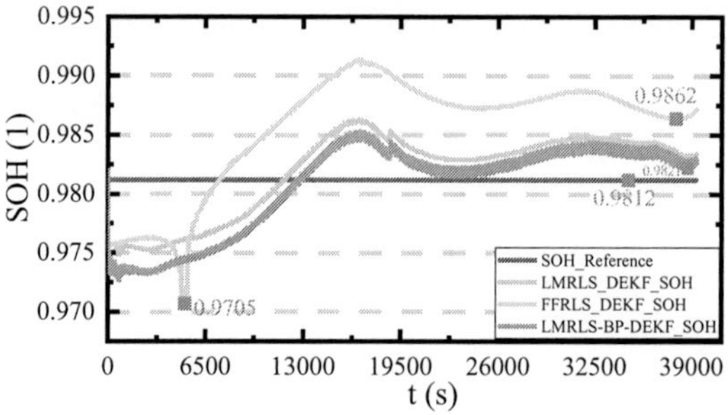

(c) Comparison of SOH estimation results of different algorithms

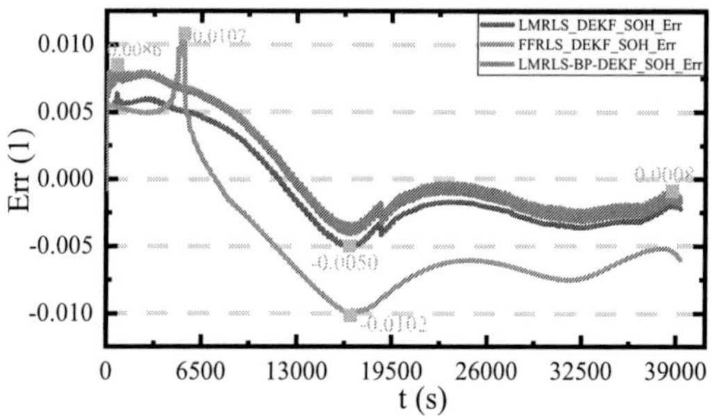

(d) Error comparison of SOH estimation results of different algorithms

Fig. 5-15 Verification results of SOC and SOH collaborative estimation of different state initial values under BBDST working conditions

From Fig. 5-15, it can be seen that the LMRLS-BPNN-DEKF collaborative estimation model proposed in this study has high accuracy and stability despite the large deviation of the initial values of SOC and SOH. As can be seen from Fig. 5-15 (a), the LMRLS-BPNN-DEKF algorithm can correct to the accurate state relatively quickly when the initial value of SOC deviates widely, and the SOC prediction curve tracks relatively well throughout the discharge process. From Fig. 5-15 (b), it can be seen that the errors in the prediction of SOC by the LMRLS-BPNN-DEKF algorithm are small and converge relatively well overall. From Fig. 5-15 (c), it can be seen that the LMRLS-BPNN-DEKF algorithm still has a relatively good tracking effect when the initial value of SOH deviates from the reference value, and its SOH prediction curve can be calibrated to the reference value and eventually converge near the reference curve. From Fig. 5-15 (d), it can be seen that the error curve of the LMRLS-BPNN-DEKF algorithm fluctuates less and the overall error value is relatively small. In summary, the algorithm model proposed in this study still has high accuracy and robustness when the initial values of SOC and SOH deviate greatly from the real values.

Table 5-8 Error indicators of SOC and SOH co-estimation for different state initial values under BBDST working conditions

Estimation method	MAXE	MAE	RMSE
FFRLS-EKF(SOC)	-1.85%	1.87%	2.03%
FFRLS-DEKF(SOC)	-1.39%	1.39%	1.79%
LMRLS-EKF(SOC)	-1.31%	1.32%	1.72%
LMRLS-DEKF(SOC)	-1.18%	1.20%	1.34%
LMRLS-BPNN-DEKF(SOC)	0.16%	0.90%	1.08%
FFRLS-DEKF(SOH)	-0.44%	0.62%	0.65%
LMRLS-DEKF(SOH)	-0.07%	0.33%	0.36%
LMRLS-BPNN-DEKF(SOH)	0.04%	0.31%	0.35%

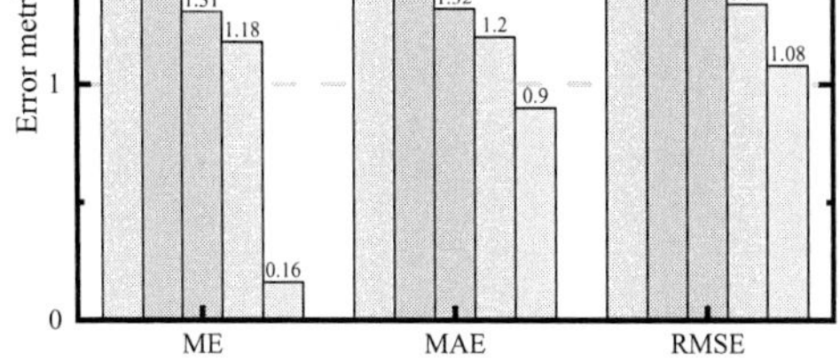

(a) Error indicators comparison of SOC estimation results

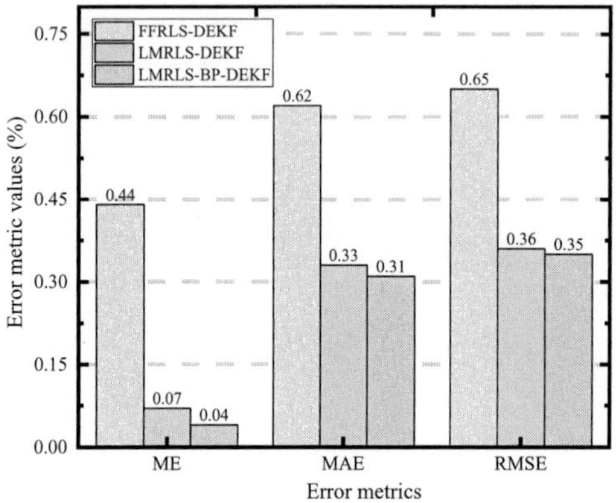

(b) Error indicators comparison of SOH estimation results

Fig. 5-16 Comparison chart of SOC/SOH co-estimation error indicators for different state initial values under BBDST working condition

From the analysis of Table 5-8 and Fig. 5-16, it is clear that the LMRLS-BPNN-DEKF collaborative estimation algorithm proposed in this study still has the best accuracy in predicting SOC value and SOH value when the initial values of SOC and SOH deviate widely. The ME of the LMRLS-BPNN-DEKF algorithm for predicting SOC and SOH is 0.16% and 0.04% respectively, which is closest to 0 compared to the other algorithms. The MAE and RMSE of SOC predicted by the algorithm are 0.90% and 1.08% respectively, which remain at about 1%, and about 0.3%~1% lower compared to the other algorithms. The MAE and RMSE of the predicted SOH are within 0.35%, which is a certain degree of improvement compared to the FFRLS-DEKF algorithm and the LMRLS-DEKF algorithm. In summary, the LMRLS-BPNN-DEKF synergistic estimation algorithm model proposed in this study has good accuracy and robustness even when the initial value of the state deviates from the true value, which indicates that the LMRLS-BPNN-DEKF algorithm has a strong correction capability and high stability.

5.4. Chapter summary

This chapter designs the experimental verification strategy of the LMRLS-BPNN-DEKF collaborative estimation algorithm model proposed in this study on the basis of Chapter 2.2. First, the verification under different temperatures and complex working conditions is engineered, and then, the model verification of the initial values of different states is further devised. To verify the effectiveness and adaptability of the LMRLS-BPNN-DEKF cooperative estimation algorithm proposed in this study, the data under DST working conditions are taken as the learning and training data of the BP neural network, and after the BP neural network reached the training target, the data under HPPC and BBDST working conditions are employed for validation analysis, respectively. The experimental results show that the LMRLS-BPNN-DEKF algorithm has high accuracy and stability for both SOC and SOH under complex operating conditions at different temperatures, and the MAE and RMSE are kept within 1%.

To further verify the robustness and correction capability of the LMRLS-BPNN-DEKF algorithm, the experimental data at 25°C are selected for experimental analysis, and the initial values of SOC and SOH are set to 0.8 and 0.9757, respectively, so that their initial states deviate from the real values. The experimental results show that the LMRLS-BPNN-DEKF cooperative estimation algorithm proposed in this study has a strong tracking ability even with large deviations of the initial state values, and the MAE and RMSE of its SOC remain around 1%, and the MAE and RMSE of SOH are stable within 0.35%. It is shown that the LMRLS-BPNN-DEKF algorithm proposed in this study has high accuracy and robustness under various complex working conditions.

CHAPTER 6

SUMMARY AND PROSPECT

6.1. Summary of the research work in this book

With the rapid development of new energy vehicles in the past two years, the power lithium battery as its power source has also been vigorously developed (Wang et al., 2023c). Due to the relatively active characteristics of lithium elements, lithium batteries are more prone to safety problems. Therefore, in order to improve the safety and endurance of lithium batteries, a reliable and effective power management system is needed to monitor them in real time (Garcia et al., 2023b). SOC and SOH are the core parameters for the battery management system to monitor the battery status, which not only relates to whether the battery management system can effectively control the battery but also are important guarantees for the electric vehicle to output reliable power and improve the battery range and driving safety (Shrivastava et al., 2023; Zhang et al., 2023c). To improve the accuracy and robustness of real-time estimation of SOC and SOH, this study proposes an LMRLS-BPNN-DEKF collaborative estimation algorithm model, that is, by designing a restricted memory recursive least squares algorithm to improve the accuracy of online parameter identification, applying a dual Kalman filter to achieve the collaborative estimation of SOC and SOH, and introducing a BP neural network to correct the model error of EKF. The main work accomplished in this book study is as follows:

(1) To understand the working characteristics of the power lithium-ion batteries and the changes in rules of its key parameters, the experimental

data are obtained through various characteristic experiments of the lithium-ion batteries to explore and analyze the hybrid pulse test characteristics, OCV characteristics at different temperatures, battery capacity characteristics at different temperatures and discharge characteristics at different rates. It provides a scientific basis for subsequent battery modeling and collaborative estimation strategies of SOC and SOH. The results of the study show that temperature has an effect on the OCV of lithium-ion batteries and there is a certain pattern, when the temperature increases (decreases), the OCV of lithium-ion batteries will also increase (decrease). The capacity change of lithium-ion batteries will continue to rise (fall) as the temperature rises (fall). When the temperature reaches a certain limit, the battery capacity tends to be stable. The change of the terminal voltage during the constant current discharge process is divided into three stages: "steep-smooth-steep". With the continuous increase of the discharge rate, the battery discharge time is gradually shortened. The larger the rate, the faster the curve drops, and the "flat period" is also shorter.

(2) To establish an accurate battery model and achieve high-precision online parameter identification, the second-order RC equivalent circuit model is finally adopted by comparing the advantages and disadvantages of different equivalent circuit models, while taking into account the model complexity and accuracy. In view of the shortcomings of the RLS algorithm and FFRLS algorithm, which have the poor anti-interference ability and are prone to oversaturation when there is too much data, this research designs the LMRLS algorithm, that is, when new data is input, old data is removed, and only the latest data with a limited length is used for parameter estimation. In this way, more information can be obtained from the new data, thereby improving the accuracy of parameter identification. To illustrate the effectiveness of the LMRLS parameter identification algorithm, the simulation experiment verification and analysis are carried out under complex working conditions of different temperatures. The results show that the LMRLS algorithm can well follow the curve of the reference

voltage under the complex working conditions of different temperatures, both MAE and RMSE remained within 2%. Therefore, the LMRLS algorithm has good convergence and tracking performance at different temperatures, with high robustness and accuracy.

(3) Considering the coupling effect between SOC and SOH, it is difficult to ensure the true state estimation value by estimating SOC/SOH alone. In this study, the ampere-hour model and capacity model of the battery are established based on the second-order model, and taking the ampere-hour integral as a bridge, the dual Kalman filter is employed to realize the progressive mutual cyclic correction strategy of the two. Among them, the first EKF realizes the estimation of SOC and uses the estimation result as the input of the second EKF to realize the progressive estimation of SOH. Then, the estimated SOH value at the current time obtained by the second EKF is used as the input of the next iteration cycle to further correct the estimated SOC value at the next time, forming a closed loop until the end of the iteration. After the mutual correction feedback of the two, the synergistic estimation is finally realized. Finally, based on the powerful self-learning ability and nonlinear mapping ability of the BP neural network, which are used for training and correction of the related data to further improve the estimation accuracy.

(4) To verify the effectiveness of the proposed LMRLS-BPNN-DEKF collaborative estimation algorithm model, an experimental test platform is established to collect experimental data of voltage and current for three complex operating conditions at 15°C, 25°C, and 35°C. After data processing and feature analysis, the experimental data set is constructed. Taking the DST working condition as the training set, after the BP neural network training reaches the predetermined goal, the HPPC and BBDST data are used for verification and analysis, and compared with different algorithms. The experimental results show that the algorithm proposed in this study has good tracking performance at various temperatures, and the prediction accuracy and stability of SOC and SOH are high.

Among them, the MAE and RMSE of the SOC predicted error by the LMRLS-BPNN-DEKF algorithm under HPPC operating conditions at 15°C are 0.65% and 1.11%, respectively, which are reduced by 0.46% to 1.74% compare to other algorithms. The MAE and RMSE for predicting SOH are 0.20% and 0.27%, respectively, which reduces by 0.16% to 0.39% compared to other algorithms. The MAE and RMSE of the SOC predicted errors by the LMRLS-BPNN-DEKF algorithm under 15°C BBDST operating conditions are 0.13% and 0.14% respectively, which are dropped by 1.37% to 1.98% compared with other algorithms. The MAE and RMSE of the predicted SOH are 0.43% and 0.60% respectively, a drop of 0.12% to 0.45% compared to other algorithms.

The MAE and RMSE of the LMRLS-BPNN-DEKF algorithm for predicting SOC errors under HPPC operating conditions at 25°C are 0.52% and 0.55%, respectively, which are descended by 0.67% to 1.62% compared to other algorithms. The MAE and RMSE for predicting SOH are 0.07% and 0.09%, respectively, with a maximum improvement of about 0.12% compared to other algorithms. The MAE and RMSE of the LMRLS-BPNN-DEKF algorithm for prognosis SOC errors under BBDST operating conditions at 25°Care 0.53% and 0.59%, respectively, which fall by 0.92% to 1.42% compared to other algorithms. The MAE and RMSE for prognosis SOH are 0.18% and 0.24%, respectively, which fall by 0.06% to 0.15% compared to other algorithms.

The MAE and RMSE of SOC errors predicted by the LMRLS-BPNN-DEKF algorithm under HPPC operating condition at 35°C are 0.13% and 0.18%, respectively, which are decreased by 0.86% to 1.90% compared with other algorithms. The MAE and RMSE of SOH are predicted to be 0.02% and 0.03% respectively, which are decreased by 0.01% to 0.09% compared with other algorithms. The MAE and RMSE of the LMRLS-BPNN-DEKF algorithm for estimating SOC errors under BBDST operating conditions at 35°C are 0.36% and 0.43%, respectively, which are diminished by 0.69% to 1.55% compared to other algorithms; The MAE and RMSE for evaluating SOH are 0.18% and 0.25%, respectively, which lessened by 0.04% to 0.57% compared to other algorithms.

To further validate and analyze the robustness and calibration performance of the LMRLS-BPNN-DEKF collaborative estimation model, the initial value of SOC is set to 0.8 and the initial value of SOH to 0.9757 at 25°C. The MAE and RMSE of its predicted SOC are 0.90% and 1.08%, respectively, which remained within 1%, with an improvement of 0.3% to 1% relative to other algorithms. Its predicted MAE and RMSE of SOH are within 0.35%, which has a certain degree of improvement compared to other algorithms.

The above research work shows that the proposed LMRLS-BPNN-DEKF synergistic estimation algorithm in this book can effectively improve the accuracy and robustness of SOC and SOH estimation and has strong correction capability and high stability, which not only lays a theoretical foundation for battery condition monitoring but also provides a reliable guarantee for the safe application of lithium batteries.

6.2. Research prospect

Aiming at the research on the high-precision collaborative estimation of SOC and SOH of power lithium-ion batteries, this study carried out a series of researches such as experimental design and characteristic analysis, algorithm design and experimental verification analysis, and realized the effective improvement of the accuracy and convergence of the of collaborative estimation of lithium battery SOC and SOH, which achieves certain research progress and results. It has certain practical significance for improving the battery management system and promoting the development of new energy vehicles, but there are still some shortcomings.

(1) Although the research on this topic considers the use of complex working conditions at different temperatures for algorithm verification, in practical applications, the temperature is not constant, and it is inevitable to

suffer from extreme temperature changes. Therefore, the authors will add temperature as an influencing factor to lithium-ion battery modeling and algorithm design in future research to build a battery model and algorithm model that is more accurate and in line with practical applications.

(2) The present study applies to the co-estimation of SOC and SOH at short time scales, and the influence of the time scale factor is not fully considered. Therefore, the authors will add the time scale factor in future studies to investigate the synergistic estimation at multiple time scales so as to further improve the adaptability of the collaborative estimation algorithm model in practical applications.

(3) This study considers the collaborative estimation of SOC and SOH of lithium battery cells, but in practical applications, it is mostly applied in the form of lithium battery packs. The state estimation of lithium battery packs needs to consider many aspects such as battery equalization and so on, which will greatly increase the difficulty and complexity of the research, and the authors will add the study on the equilibrium strategy of lithium battery packs in the follow-up research work.

REFERENCES

Adaikkappan, M., Sathiyamoorthy, N., 2022. Modeling, state of charge estimation, and charging of lithium-ion battery in electric vehicle: A review. International Journal of Energy Research 46 (3), 2141-2165. https://doi.org/10.1002/er.7339.

Ahmed, A.A., Nazzal, M.A., Darras, B.M., Deiab, I.M., 2023. Global warming potential, water footprint, and energy demand of shared autonomous electric vehicles incorporating circular economy practices. Sustainable Production and Consumption 36, 449-462. https://doi.org/10.1016/j.spc.2023.02.001.

Bao, W.K., Liu, H.D., Sun, Y.D., Zheng, Y.J., 2022a. A Fast Prediction of Open-Circuit Voltage and a Capacity Estimation Method of a Lithium-Ion Battery Based on a BP Neural Network. Batteries-Basel 8 (12), 1-18. https://doi.org/10.3390/batteries8120289.

Bao, W.K., Liu, H.D., Sun, Y.D., Zheng, Y.J., 2022b. A Fast Prediction of Open-Circuit Voltage and a Capacity Estimation Method of a Lithium-Ion Battery Based on a BP Neural Network. Batteries-Basel 8 (12). https://doi.org/10.3390/batteries8120289.

Bobobee, E.D., Wang, S.L., Zou, C.Y., Takyi-Aninakwa, P., Zhou, H., Appiah, E., 2023. State of charge estimation of ternary lithium-ion batteries at variable ambient temperatures. International Journal of Electrochemical Science 18 (4). https://doi.org/10.1016/j.ijoes.2023.100062.

Cai, L., Lin, J.D., Liao, X.Y., 2022. A data-driven method for state of health prediction of lithium-ion batteries in a unified framework. Journal of

Energy Storage 51. https://doi.org/10.1016/j.est.2022.104371.

Cao, J., Wang, S.L., Xie, Y.X., Fernandez, C., 2023. Research on the state of charge estimation method of lithium-ion batteries based on novel limited memory multi-innovation least squares method and SDE-2-RC equivalent model. International Journal of Circuit Theory and Applications 51 (4), 1902-1917. https://doi.org/10.1002/cta.3500.

Chang, C., Wang, Q.Y., Jiang, J.C., Wu, T.Z., 2021. Lithium-ion battery state of health estimation using the incremental capacity and wavelet neural networks with genetic algorithm. Journal of Energy Storage 38, 1-9. https://doi.org/10.1016/j.est.2021.102570.

Chemali, E., Kollmeyer, P.J., Preindl, M., Emadi, A., 2018. State-of-charge estimation of Li-ion batteries using deep neural networks: A machine learning approach. Journal of Power Sources 400, 242-255. https://doi.org/10.1016/j.jpowsour.2018.06.104.

Chen, D.D., Xiao, L., Yan, W.D., Guo, Y.B.A., 2021a. A novel hybrid equivalent circuit model for lithium-ion battery considering nonlinear capacity effects. Energy Reports 7, 320-329. https://doi.org/10.1016/j.egyr.2021.06.051.

Chen, L., Shunli, W., Jiang, H., Fernandez, C., 2022. An adaptive fractional-order unscented Kalman filter for Li-ion batteries in the energy storage system. Indian Journal of Physics 96 (13), 3933-3939. https://doi.org/10.1007/s12648-022-02314-2.

Chen, L., Wang, S.L., Jiang, H., Fernandez, C., 2023. A novel combined estimation method for state of energy and predicted maximum available energy based on fractional-order modeling. Journal of Energy Storage 62. https://doi.org/10.1016/j.est.2023.106930.

Chen, L., Wang, S.L., Jiang, H., Fernandez, C., Xiong, X., 2021b. A novel combined estimation method of online full-parameter identification and adaptive unscented particle filter for Li-ion batteries SOC based on fractional-order modeling. International Journal of Energy Research 45 (10), 15481-15494. https://doi.org/10.1002/er.6817.

Chen, L., Wang, S.L., Jiang, H., Fernandez, C., Xiong, X., 2021c. A Novel Fractional – Order Extended Kalman Filtering Method for on-line Joint State Estimation and Parameter Identification of the High Power Li-ion Batteries. International Journal of Electrochemical Science 16 (5). https://doi.org/10.20964/2021.05.64.

Chen, X.P., Wang, S.L., Xie, Y.X., Fernandez, C., Fan, Y.C., 2021d. A novel Fireworks Factor and Improved Elite Strategy based on Back Propagation Neural Networks for state-of-charge estimation of lithium-ion batteries. International Journal of Electrochemical Science 16 (9). https://doi.org/10.20964/2021.08.07.

Cheng, G., Zhao, C., Iqbal, N., Gulmez, O., Isik, H., Kirikkaleli, D., 2021. Does energy productivity and public-private investment in energy achieve carbon neutrality target of China? Journal of Environmental Management 298. https://doi.org/10.1016/j.jenvman.2021.113464.

Cui, Z.H., Wang, L.C., Li, Q., Wang, K., 2022. A comprehensive review on the state of charge estimation for lithium-ion battery based on neural network. International Journal of Energy Research 46 (5), 5423-5440. https://doi.org/10.1002/er.7545.

Deng, J., Zhang, Y., Xing, X., Liu, C., 2022. Can Carbon Neutrality Commitment Contribute to the Sustainable Development of China's New Energy Companies? Sustainability 14 (18). https://doi.org/10.3390/su141811308.

Du, B., Lund, P.D., Wang, J., 2022. Improving the accuracy of predicting the performance of solar collectors through clustering analysis with artificial neural network models. Energy Reports 8, 3970-3981. https://doi.org/10.1016/j.egyr.2022.03.013.

Du, C.Y., Zhao, Z.Y., Liu, H., Song, F.Y., Chen, L.L., Cheng, Y., Guo, Z.H., 2023a. The Status of Representative Anode Materials for Lithium-Ion Batteries. Chemical Record. https://doi.org/10.1002/tcr.202300004.

Du, X.H., Meng, J.H., Liu, K.L., Zhang, Y.M., Wang, S.L., Peng, J.C., Liu, T.Q., 2023b. Online Identification of Lithium-ion Battery Model

Parameters with Initial Value Uncertainty and Measurement Noise. Chinese Journal of Mechanical Engineering 36 (1), 1-10. https://doi.org/10.1186/s10033-023-00846-0.

Du, X.H., Meng, J.H., Liu, K.L., Zhang, Y.M., Wang, S.L., Peng, J.C., Liu, T.Q., 2023c. Online Identification of Lithium-ion Battery Model Parameters with Initial Value Uncertainty and Measurement Noise. Chinese Journal of Mechanical Engineering 36 (1). https://doi.org/10.1186/s10033-023-00846-0.

Duan, J.D., Wang, P., Ma, W.T., Qiu, X.Y., Tian, X., Fang, S., 2020. State of Charge Estimation of Lithium Battery Based on Improved Correntropy Extended Kalman Filter. Energies 13 (16), 1-18. https://doi.org/10.3390/en13164197.

Fan, Y.C., Qiu, J.S., Wang, S.L., Yang, X., Liu, D.L., Fernandez, C., 2022. Incremental Capacity Curve Health-Indicator Extraction Based on Gaussian Filter and Improved Relevance Vector Machine for Lithium-Ion Battery Remaining Useful Life Estimation. Metals 12 (8). https://doi.org/10.3390/met12081331.

Fan, Y.C., Shi, H.T., Wang, S.L., Fernandez, C., Cao, W., Huang, J.H., 2021a. A Novel Adaptive Function-Dual Kalman Filtering Strategy for Online Battery Model Parameters and State of Charge Co-Estimation. Energies 14 (8). https://doi.org/10.3390/en14082268.

Fan, Y.C., Wang, S.L., Jiang, C., Fernandez, C., 2021b. The Power State Estimation Method for High Energy Ternary Lithium-ion Batteries Based on the Online Collaborative Equivalent Modeling and Adaptive Correction – Unscented Kalman Filter. International Journal of Electrochemical Science 16 (1). https://doi.org/10.20964/2021.01.70.

Gao, H.Y., Wang, S.L., Qiao, J.L., Yang, X., Fernandez, C., 2022a. An Improved Weighting Coefficient Optimization-Particle Filtering Algorithm Based on Gaussian Degradation Model for Remaining Useful Life Prediction of Lithium-ion Batteries. Journal of the Electrochemical Society 169 (12). https://doi.org/10.1149/1945-

7111/aca6a2.

Gao, Z., Rao, S., Zhang, T., Li, W., Yang, X., Chen, Y., Zheng, Y., Ding, Y., Dong, T., Li, S., 2022b. Design Strategies of Flame-Retardant Additives for Lithium Ion Electrolyte. Journal of Electrochemical Energy Conversion and Storage 19 (3). https://doi.org/10.1115/1.4053968.

Garcia, A., Monsalve-Serrano, J., Martinez-Boggio, S., Golke, D., 2023a. Energy assessment of the ageing phenomenon in Li-Ion batteries and its impact on the vehicle range efficiency. Energy Conversion and Management 276, 1-25. https://doi.org/10.1016/j.enconman.2022.116530.

Garcia, E., Quiles, E., Correcher, A., 2023b. Distributed Intelligent Battery Management System Using a Real-World Cloud Computing System. Sensors (Basel, Switzerland) 23 (7). https://doi.org/10.3390/s23073417.

Gong, Q., Wang, P., Cheng, Z., 2022. A Data-Driven Model Framework Based on Deep Learning for Estimating the States of Lithium-Ion Batteries. Journal of the Electrochemical Society 169 (3), 1-10. https://doi.org/10.1149/1945-7111/ac5bac.

He, M.F., Wang, S.L., Fernandez, C., Vu, C.M., Li, X.X., Bobobee, E.D., 2021. A Novel Adaptive Particle Swarm Optimization Algorithm Based High Precision Parameter Identification and State Estimation of Lithium-Ion Battery. International Journal of Electrochemical Science 16 (5). https://doi.org/10.20964/2021.05.55.

Hong, J.C., Wang, Z.P., Chen, W., Wang, L.Y., Qu, C.H., 2020. Online joint-prediction of multi-forward-step battery SOC using LSTM neural networks and multiple linear regression for real-world electric vehicles. Journal of Energy Storage 30, 1-12. https://doi.org/10.1016/j.est.2020.101459.

Hong, S., Yue, T.Y., Liu, H., Vehicle energy system active defense: A health assessment of lithium-ion batteries. International Journal of

Intelligent Systems. https://doi.org/10.1002/int.22309.

Hong, Y.H., Ma, Z., Li, K.X., Li, J.Y., Tang, S.Y., Xu, Z., Yu, D.D., Chen, D., Qin, L.S., Xie, J., He, Q.G., 2023. Research progress and perspectives on ultra-low temperature organic batteries. Journal of Materials Chemistry A 11 (15), 7898-7923. https://doi.org/10.1039/d2ta09988h.

Hossain, M., Haque, M.E., Arif, M.T., 2022. Kalman filtering techniques for the online model parameters and state of charge estimation of the Li-ion batteries: A comparative analysis. Journal of Energy Storage 51, 1-24. https://doi.org/10.1016/j.est.2022.104174.

Hu, S.M., Liu, Z.H., Tan, Y.S., Cheng, X., Chen, Z.J., Long, Z.M., 2022a. The status quo and future trends of new energy vehicle power batteries in China - Analysis from policy perspective. Energy Reports 8, 63-80. https://doi.org/10.1016/j.egyr.2022.09.082.

Hu, Y.F., Wangz, S.L., Huang, J.H., Takyi-Aninakwa, P., Chen, X.P., 2022b. A Novel Seasonal Autoregressive Integrated Moving Average Method for the Accurate Lithium-ion Battery Residual Life Prediction. International Journal of Electrochemical Science 17 (5). https://doi.org/10.20964/2022.05.61.

Huang, J.H., Wang, S.L., Xu, W.H., Fernandez, C., Fan, Y.C., Chen, X.P., 2021a. An Improved Rainflow Algorithm Combined with Linear Criterion for the Accurate Li-ion Battery Residual Life Prediction. International Journal of Electrochemical Science 16 (7). https://doi.org/10.20964/2021.07.29.

Huang, J.H., Wang, S.L., Xu, W.H., Shi, W.H., Fernandez, C., 2021b. A Novel Autoregressive Rainflow-Integrated Moving Average Modeling Method for the Accurate State of Health Prediction of Lithium-Ion Batteries. Processes 9 (5). https://doi.org/10.3390/pr9050795.

Huang, W.-Z., Zhao, C.-Z., Wu, P., Yuan, H., Feng, W.-E., Liu, Z.-Y., Lu, Y., Sun, S., Fu, Z.-H., Hu, J.-K., Yang, S.-J., Huang, J.-Q., Zhang, Q., 2022. Anode-Free Solid-State Lithium Batteries: A Review. Advanced

Energy Materials 12 (26). https://doi.org/10.1002/aenm.202201044.
Huang, W., Feng, X., Han, X., Zhang, W., Jiang, F., 2021c. Questions and Answers Relating to Lithium-Ion Battery Safety Issues. Cell Reports Physical Science 2 (1). https://doi.org/10.1016/j.xcrp.2020.100285.
Ji, W.K., Wang, S.L., Zou, C.A.Y., Shi, H.T., 2021. A Novel Fading Memory Square Root UKF Algorithm for the High-precision State of Charge Estimation of High-power Lithium-ion Batteries. International Journal of Electrochemical Science 16 (7). https://doi.org/10.20964/2021.07.68.
Jia, X.Y., Wang, S.L., Qiao, J.L., Cao, W., 2022. An adaptive spherical square-root double unscented Kalman filtering algorithm for estimating state-of-charge of lithium-ion batteries. International Journal of Energy Research 46 (10), 14256-14267. https://doi.org/10.1002/er.8139.
Jiang, C., Wang, S.L., Wu, B., Fernandez, C., Xiong, X., Coffie-Ken, J., 2021. A state-of-charge estimation method of the power lithium-ion battery in complex conditions based on adaptive square root extended Kalman filter. Energy 219. https://doi.org/10.1016/j.energy.2020.119603.
Jiao, M., Wang, D.Q., Qiu, J.L., 2020. A GRU-RNN based momentum optimized algorithm for SOC estimation. Journal of Power Sources 459, 1-10. https://doi.org/10.1016/j.jpowsour.2020.228051.
Jin, S.Y., Sui, X., Huang, X.R., Wang, S.L., Teodorescu, R., Stroe, D.I., 2021a. Overview of Machine Learning Methods for Lithium-Ion Battery Remaining Useful Lifetime Prediction. Electronics 10 (24). https://doi.org/10.3390/electronics10243126.
Jin, X.B., Jeremiah, R.J.R., Su, T.L., Bai, Y.T., Kong, J.L., 2021b. The New Trend of State Estimation: From Model-Driven to Hybrid-Driven Methods. Sensors 21 (6). https://doi.org/10.3390/s21062085.
Jo, S., Jung, S., Roh, T., 2021. Battery State-of-Health Estimation Using Machine Learning and Preprocessing with Relative State-of-Charge.

Energies 14 (21). https://doi.org/10.3390/en14217206.

Khodaparast, J., 2022. A Review of Dynamic Phasor Estimation by Non-Linear Kalman Filters. Ieee Access 10, 11090-11109. https://doi.org/10.1109/access.2022.3146732.

Kong, D.P., Wang, S.H., Ping, P., 2021a. A novel parameter adaptive method for state of charge estimation of aged lithium batteries. Journal of Energy Storage 44, 1-13. https://doi.org/10.1016/j.est.2021.103389.

Kong, F., 2022. A better understanding of the role of new energy and green finance to help achieve carbon neutrality goals, with special reference to China. Science Progress 105 (1), 1-23. https://doi.org/10.1177/00368504221086361.

Kong, X.Y., Zhang, X.P., Lu, N., Ma, Y.Y., Li, Y., 2021b. Online Smart Meter Measurement Error Estimation Based on EKF and LMRLS Method. Ieee Transactions on Smart Grid 12 (5), 4269-4279. https://doi.org/10.1109/tsg.2021.3077693.

Lai, X., Qin, C., Gao, W.K., Zheng, Y.J., Yi, W., 2018. A State of Charge Estimator Based Extended Kalman Filter Using an Electrochemistry-Based Equivalent Circuit Model for Lithium-Ion Batteries. Applied Sciences-Basel 8 (9), 1-15. https://doi.org/10.3390/app8091592.

Li, B.W., Wang, S.L., Fernandez, C., Yu, C.M., Shi, W.H., Fan, Y.C., 2021a. A Linear Recursive State of Power Estimation for Fusion Model Component Analysis with Constant Sampling Time. International Journal of Electrochemical Science 16 (9). https://doi.org/10.20964/2021.09.14.

Li, B.W., Wang, S.L., Fernandez, C., Yu, C.M., Xia, L.L., Fan, Y.C., 2021b. A linear recursive state of power estimation method based on fusion model of voltage and state of charge limitations. Journal of Energy Storage 40. https://doi.org/10.1016/j.est.2021.102583.

Li, H., Zou, C.Y., Fernandez, C., Wang, S.L., Fan, Y.C., Liu, D.L., 2021c. A Novel State of Charge Estimation for Energy Storage Systems Based on the Joint NARX Network and Filter Algorithm. International

Journal of Electrochemical Science 16 (12). https://doi.org/10.20964/2021.12.50.

Li, J., Wang, S.L., Chen, L., He, M.F., 2022a. Adaptive Exponentially Weighted Extended Kalman Filtering for State of Charge Estimation of Lithium-ion Battery. International Journal of Electrochemical Science 17 (7). https://doi.org/10.20964/2022.07.63.

Li, K., Su, Q., Ma, X., Zhang, H., 2023a. Research on Lithium Technology Safety Issues: A Bibliometric Analysis. Sustainability 15 (5). https://doi.org/10.3390/su15054128.

Li, Y., Wang, S.L., Chen, L., Yu, P., Chen, X.P., 2022b. Research on state-of-charge Estimation of Lithium-ion Batteries Based on Improved Sparrow Search Algorithm-BP Neural Network. International Journal of Electrochemical Science 17 (8). https://doi.org/10.20964/2022.08.48.

Li, Z., Shen, S., Zhou, Z., Cai, Z., Gu, W., Zhang, F., 2023b. Novel method for modelling and adaptive estimation for SOC and SOH of lithium-ion batteries. Journal of Energy Storage 62, 1-14. https://doi.org/10.1016/j.est.2023.106927.

Lin, H.Y., Kang, L.Y., Xie, D., Linghu, J.Q., Li, J., 2022. Online State-of-Health Estimation of Lithium-Ion Battery Based on Incremental Capacity Curve and BP Neural Network. Batteries-Basel 8 (4). https://doi.org/10.3390/batteries8040029.

Lin, M., Wu, J., Meng, J., Wang, W., Wu, J., 2023a. State of health estimation with attentional long short-term memory network for lithium-ion batteries. Energy 268. https://doi.org/10.1016/j.energy.2023.126706.

Lin, P., Jin, P., Zhang, H.Y., 2023b. Identification and Fast Measurement Method of Open-circuit Voltage. Journal of the Electrochemical Society 170 (3). https://doi.org/10.1149/1945-7111/acc2ec.

Lipu, M.S.H., Hannan, M.A., Hussain, A., Hoque, M.M., Ker, P.J., Saad, M.H.M., Ayob, A., 2018. A review of state of health and remaining useful life estimation methods for lithium-ion battery in electric vehicles: Challenges and recommendations. Journal of Cleaner Production 205, 115-133. https://doi.org/10.1016/j.jclepro.2018.09.065.

Liu, B.C., Song, C.Y., Wang, Q.S., Zhang, X.M., Chen, J.L., 2022a. Research on regional differences of China's new energy vehicles promotion policies: A perspective of sales volume forecasting. Energy 248. https://doi.org/10.1016/j.energy.2022.123541.

Liu, D.L., Fan, Y.C., Wang, S.L., Xia, L.L., Qiu, J.S., Bobobee, E.D., 2021a. A Novel Fading Memory Recursive Least Square Method (FMLS) for Accurate State of Charge Estimation of Lithiumion Batteries Combined with Improved Second Order PNGV Modeling. International Journal of Electrochemical Science 16 (9). https://doi.org/10.20964/2021.09.34.

Liu, D.L., Wang, S.L., Fan, Y.C., Xia, L.L., Qiu, J.S., 2022b. A novel fuzzy-extended Kalman filter-ampere-hour (F-EKF-Ah) algorithm based on improved second-order PNGV model to estimate state of charge of lithium-ion batteries. International Journal of Circuit Theory and Applications 50 (11), 3811-3826. https://doi.org/10.1002/cta.3386.

Liu, H., Hu, F., Su, J.S., Wei, X.W., Qin, R.S., 2020. Comparisons on Kalman-Filter-Based Dynamic State Estimation Algorithms of Power Systems. Ieee Access 8, 51035-51043. https://doi.org/10.1109/access.2020.2979735.

Liu, J.L., Duan, Q.L., Qi, K.X., Liu, Y.J., Sun, J.H., Wang, Z.R., Wang, Q.S., 2022c. Capacity fading mechanisms and state of health prediction of commercial lithium-ion battery in total lifespan. Journal of Energy Storage 46. https://doi.org/10.1016/j.est.2021.103910.

Liu, J.W., Zheng, M.X., Wu, S.Y., Zhang, L., 2023a. Design strategies for coordination polymers as electrodes and electrolytes in rechargeable lithium batteries. Coordination Chemistry Reviews 483. https://doi.org/10.1016/j.ccr.2023.215084.

Liu, S.-X., Zhou, Y.-F., Liu, Y.-L., Lian, J., Huang, L.-J., 2021b. A Method for Battery Health Estimation Based on Charging Time Segment. Energies 14 (9). https://doi.org/10.3390/en14092612.

Liu, S.Z., Chen, J.J., Zhang, C., Jin, L., Yang, Q.X., 2022d. Experimental study on lithium-ion cell characteristics at different discharge rates. Journal of Energy Storage 45. https://doi.org/10.1016/j.est.2021.103418.

Liu, X.T., Li, Q.L., Wang, L., Lin, M.Q., Wu, J., 2023b. Data-Driven State of Charge Estimation for Power Battery With Improved Extended Kalman Filter. Ieee Transactions on Instrumentation and Measurement 72. https://doi.org/10.1109/tim.2023.3239629.

Liu, Z.H., Wang, C., Guo, X.M., Cheng, S.K., Gao, Y.H., Wang, R., Sun, Y.H., Yan, P., 2021c. Thermal characteristics of ultrahigh power density lithium-ion battery. Journal of Power Sources 506. https://doi.org/10.1016/j.jpowsour.2021.230205.

Long, T., Wang, S.L., Cao, W., Ren, P., He, M.F., Fernandez, C., 2022. Collaborative state estimation of lithium-ion battery based on multi-time scale low-pass filter forgetting factor recursive least squares - double extended Kalman filtering algorithm. International Journal of Circuit Theory and Applications 50 (6), 2108-2127. https://doi.org/10.1002/cta.3250.

Long, T., Wang, S.L., Cao, W., Zhou, H., Fernandez, C., 2023. An improved variable forgetting factor recursive least square-double extend Kalman filtering based on global mean particle swarm optimization algorithm for collaborative state of energy and state of health estimation of lithium-ion batteries. Electrochimica Acta 450. https://doi.org/10.1016/j.electacta.2023.142270.

Lu, X., Tong, D., He, K., 2023a. China's carbon neutrality: an extensive and profound systemic reform. Frontiers of Environmental Science & Engineering 17 (2). https://doi.org/10.1007/s11783-023-1614-3.

Lu, X.Y., Li, Z.R., Zou, J.T., Peng, D.Z., Hu, W.Y., Zhong, Y.Y., Li, P.F., Su, S.L., Wang, X.W., Wang, S.B., Zhao, Z.W., Zhang, J.F., 2023b. Spent lithium manganate batteries for sustainable recycling: A review. Frontiers in Materials 10. https://doi.org/10.3389/fmats.2023.1152018.

Ma, K.Z., Kong, J., Wang, Y.H., Lu, X.M., 2022a. Review of the Applications of Kalman Filtering in Quantum Systems. Symmetry-Basel 14 (12). https://doi.org/10.3390/sym14122478.

Ma, L., Xu, Y., Zhang, H., Yang, F., Wang, X., Li, C., 2022b. Co-estimation of state of charge and state of health for lithium-ion batteries based on fractional-order model with multi-innovations unscented Kalman filter method. Journal of Energy Storage 52. https://doi.org/10.1016/j.est.2022.104904.

Mahony, R., van Goor, P., Hamel, T., 2022. Observer Design for Nonlinear Systems with Equivariance. Annual Review of Control Robotics and Autonomous Systems 5, 221-252. https://doi.org/10.1146/annurev-control-061520-010324.

McCarthy, K., Gullapalli, H., Ryan, K.M., Kennedy, T., 2022. Electrochemical impedance correlation analysis for the estimation of Li-ion battery state of charge, state of health and internal temperature. Journal of Energy Storage 50, 1-10. https://doi.org/10.1016/j.est.2022.104608.

Meng, J.H., Luo, G.Z., Ricco, M., Swierczynski, M., Stroe, D.I., Teodorescu, R., 2018. Overview of Lithium-Ion Battery Modeling Methods for State-of-Charge Estimation in Electrical Vehicles. Applied Sciences-Basel 8 (5), 1-17. https://doi.org/10.3390/app8050659.

Miao, H., Chen, J., Mao, L., Qu, K., Zhao, J., Zhu, Y., 2021. A novel online model parameters identification method with anti-interference characteristics for lithium-ion batteries. International Journal of Energy Research 45 (6), 9502-9517. https://doi.org/10.1002/er.6477.

Naseri, F., Schaltz, E., Stroe, D.I., Gismero, A., Farjah, E., 2022. An Enhanced Equivalent Circuit Model With Real-Time Parameter Identification for Battery State-of-Charge Estimation. Ieee Transactions on Industrial Electronics 69 (4), 3743-3751. https://doi.org/10.1109/tie.2021.3071679.

Natella, D., Onori, S., Vasca, F., 2023. A Co-Estimation Framework for State of Charge and Parameters of Lithium-Ion Battery With Robustness to Aging and Usage Conditions. Ieee Transactions on Industrial Electronics 70 (6), 5760-5770. https://doi.org/10.1109/tie.2022.3194576.

Oyewole, I., Chehade, A., Kim, Y., 2022. A controllable deep transfer learning network with multiple domain adaptation for battery state-of-charge estimation. Applied Energy 312, 1-15. https://doi.org/10.1016/j.apenergy.2022.118726.

Park, J., Kim, K., Park, S., Baek, J., Kim, J., 2021. Complementary cooperative SOC/capacity estimator based on the discrete variational derivative combined with the DEKF for electric power applications. Energy 232, 1-13. https://doi.org/10.1016/j.energy.2021.121023.

Park, J., Lee, B., Jung, D.Y., Kim, D.H., 2018. Battery State Estimation Algorithm for High-Capacity Lithium Secondary Battery for EVs Considering Temperature Change Characteristics. Journal of Electrical Engineering & Technology 13 (5), 1927-1934. https://doi.org/10.5370/jeet.2018.13.5.1927.

Park, J., Lee, M., Kim, G., Park, S., Kim, J., 2020. Integrated Approach Based on Dual Extended Kalman Filter and Multivariate Autoregressive Model for Predicting Battery Capacity Using Health Indicator and SOC/SOH. Energies 13 (9), 1-20. https://doi.org/10.3390/en13092138.

Peng, J.W., Shi, H.T., Wang, S.L., Wang, L.P., Fernandez, C., Xiong, X., Dablu, B.E., 2022. A novel equivalent modeling method combined with the splice-electrochemical polarization model and prior generalized inverse least-square parameter identification for UAV lithium-ion batteries. Energy Science & Engineering 10 (10), 3727-3740. https://doi.org/10.1002/ese3.1268.

Potrc, S., Cucek, L., Martin, M., Kravanja, Z., 2021. Sustainable renewable energy supply networks optimization-The gradual transition to a renewable energy system within the European Union by 2050. Renewable & Sustainable Energy Reviews 146, 1-16. https://doi.org/10.1016/j.rser.2021.111186.

Pu, R., Wang, S.L., Chen, X.P., Huang, J.H., He, M.F., Cao, W., 2022. A novel cuckoo search particle filtering strategy for the remaining useful life prediction of the lithium-ion batteries in hybrid electric vehicle. International Journal of Energy Research 46 (15), 21703-21715. https://doi.org/10.1002/er.8712.

Qays, M.O., Buswig, Y., Hossain, M.L., Abu-Siada, A., 2022. Recent Progress and Future Trends on the State of Charge Estimation Methods to Improve Battery-storage Efficiency: A Review. Csee Journal of Power and Energy Systems 8 (1), 105-114. https://doi.org/10.17775/cseejpes.2019.03060.

Qi, C.S., Wang, S.L., Cao, W., Shi, H.T., Xie, Y.X., 2022a. A Novel Multi-Constraint Peak Power Prediction Method Combined with Online Model Parameter Identification and State-of-Charge Co-Estimation of Lithium-Ion Batteries. Journal of the Electrochemical Society 169 (12). https://doi.org/10.1149/1945-7111/aca721.

Qi, C.S., Wang, S.L., Cao, W., Yu, P., Xie, Y.X., 2022b. An improved adaptive spherical unscented Kalman filtering method for the accurate state-of-charge estimation of lithium-ion batteries. International Journal of Circuit Theory and Applications 50 (10), 3487-3502. https://doi.org/10.1002/cta.3356.

Qian, K.F., Liu, X.T., 2021. Hybrid optimization strategy for lithium-ion battery's State of Charge/Health using joint of dual Kalman filter and Modified Sine-cosine Algorithm. Journal of Energy Storage 44, 1-15.

https://doi.org/10.1016/j.est.2021.103319.

Qiao, J.L., Wang, S.L., Yu, C.M., Shi, W.H., Fernandez, C., 2021. A novel bias compensation recursive least square-multiple weighted dual extended Kalman filtering method for accurate state-of-charge and state-of-health co-estimation of lithium-ion batteries. International Journal of Circuit Theory and Applications 49 (11), 3879-3893. https://doi.org/10.1002/cta.3115.

Qiao, J.L., Wang, S.L., Yu, C.M., Yang, X., Fernandez, C., 2022. A novel intelligent weight decreasing firefly-particle filtering method for accurate state-of-charge estimation of lithium-ion batteries. International Journal of Energy Research 46 (5), 6613-6622. https://doi.org/10.1002/er.7596.

Qiao, J.L., Wang, S.L., Yu, C.M., Yang, X., Fernandez, C., 2023. A chaotic firefly- Particle filtering method of dynamic migration modeling for the state-of-charge and state-of-health co-estimation of a lithium-ion battery performance. Energy 263. https://doi.org/10.1016/j.energy.2022.126164.

Qiu, J.S., Fan, Y.C., Wang, S.L., Yang, X., Qiao, J.L., Liu, D.L., 2022. Research on the remaining useful life prediction method of lithium-ion batteries based on aging feature extraction and multi-kernel relevance vector machine optimization model. International Journal of Energy Research 46 (10), 13931-13946. https://doi.org/10.1002/er.8110.

Qiu, X.H., Wu, W.X., Wang, S.F., 2020. Remaining useful life prediction of lithium-ion battery based on improved cuckoo search particle filter and a novel state of charge estimation method. Journal of Power Sources 450. https://doi.org/10.1016/j.jpowsour.2020.227700.

Qu, X., Zhang, B.L., Zhao, J.J., Qiu, B.L., Chen, X., Zhou, F.Y., Li, X.Y., Gao, S.B., Wang, D.H., Yin, H.Y., 2023. Salt-thermal methods for recycling and regenerating spent lithium-ion batteries: a review. Green Chemistry 25 (8), 2992-3015. https://doi.org/10.1039/d2gc04620b.

Ren, P., Wang, S.L., Chen, X.P., Huang, J.H., He, M.F., 2022a. Fusion estimation strategy based on dual adaptive Kalman filtering algorithm for the state of charge and state of health of hybrid electric vehicle Li-ion batteries. International Journal of Energy Research 46 (6), 7374-7388. https://doi.org/10.1002/er.7643.

Ren, P., Wang, S.L., Chen, X.P., Zhou, H., Fernandez, C., Stroe, D.I., 2022b. A novel multiple training-scale dynamic adaptive cuckoo search optimized long short-term memory neural network and multi-dimensional health indicators acquisition strategy for whole life cycle health evaluation of lithium-ion batteries. Electrochimica Acta 435. https://doi.org/10.1016/j.electacta.2022.141404.

Ren, P., Wang, S.L., Huang, J.H., Chen, X.P., He, M.F., Cao, W., 2022c. Novel co-estimation strategy based on forgetting factor dual particle filter algorithm for the state of charge and state of health of the lithium-ion battery. International Journal of Energy Research 46 (2), 1094-1107. https://doi.org/10.1002/er.7230.

Rossi, C., Falcomer, C., Biondani, L., Pontara, D., 2022. Genetically Optimized Extended Kalman Filter for State of Health Estimation Based on Li-Ion Batteries Parameters. Energies 15 (9). https://doi.org/10.3390/en15093404.

Shan, H.P., Cao, H., Xu, X., Xiao, T., Hou, G.Y., Cao, H.Z., Tang, Y.P., Zheng, G.Q., 2022. Investigation of self-discharge properties and a new concept of open-circuit voltage drop rate in lithium-ion batteries. Journal of Solid State Electrochemistry 26 (1), 163-170. https://doi.org/10.1007/s10008-021-05049-y.

Shen, S.Q., Liu, B.C., Zhang, K., Ci, S., 2021. Toward Fast and Accurate SOH Prediction for Lithium-Ion Batteries. Ieee Transactions on Energy Conversion 36 (3), 2036-2046. https://doi.org/10.1109/tec.2021.3052504.

Shi, H.T., Wang, L.P., Wang, S.L., Fernandez, C., Xiong, X., Dablu, B.E., Xu, W.H., 2022a. A novel lumped thermal characteristic modeling strategy for the online adaptive temperature and parameter co-estimation of vehicle lithium-ion batteries. Journal of Energy Storage 50. https://doi.org/10.1016/j.est.2022.104309.

Shi, H.T., Wang, S.L., Fernandez, C., Huang, J.H., Xu, W.H., Wang, L.P., 2022b. Battery hysteresis compensation modeling and state-of-charge estimation adaptive to time-varying ambient temperature conditions. International Journal of Energy Research 46 (12), 17096-17112. https://doi.org/10.1002/er.8373.

Shi, H.T., Wang, S.L., Fernandez, C., Yu, C.M., Fan, Y.C., Cao, W., 2021a. Improved splice-electrochemical circuit polarization modeling and optimized dynamic functional multi-innovation least square parameter identification for lithium-ion batteries. International Journal of Energy Research 45 (10), 15323-15337. https://doi.org/10.1002/er.6807.

Shi, H.T., Wang, S.L., Fernandez, C., Yu, C.M., Li, X.X., Zou, C.Y., 2020. Adaptive iterative working state prediction based on the double unscented transformation and dynamic functioning for unmanned aerial vehicle lithium-ion batteries. Measurement & Control 53 (9-10), 1760-1773. https://doi.org/10.1177/0020294020923057.

Shi, H.T., Wang, S.L., Fernandez, C., Yu, C.M., Xu, W.H., Dablu, B.E., Wang, L.P., 2022c. Improved multi-time scale lumped thermoelectric coupling modeling and parameter dispersion evaluation of lithium-ion batteries. Applied Energy 324. https://doi.org/10.1016/j.apenergy.2022.119789.

Shi, H.T., Wang, S.L., Wang, L.P., Xu, W.H., Fernandez, C., Dablu, B.E., Zhang, Y.C., 2022d. On-line adaptive asynchronous parameter identification of lumped electrical characteristic model for vehicle lithium-ion battery considering multi-time scale effects. Journal of Power Sources 517. https://doi.org/10.1016/j.jpowsour.2021.230725.

Shi, W.H., Wang, S.L., Xia, L.L., Yu, P., Li, B.W., 2021b. A novel voltage matching-adaptive extended Kalman filtering construction method for the state of charge prediction of lithium-ion batteries. Transactions of the Institute of Measurement and Control. https://doi.org/10.1177/01423312211029489.

Shrivastava, P., Naidu, P.A., Sharma, S., Panigrahi, B.K., Garg, A., 2023. Review on technological advancement of lithium-ion battery states estimation methods for electric vehicle applications. Journal of Energy Storage 64. https://doi.org/10.1016/j.est.2023.107159.

Shrivastava, P., Soon, T.K., Bin Idris, M.Y.I., Mekhilef, S., 2019. Overview of model-based online state-of-charge estimation using Kalman filter family for lithium-ion batteries. Renewable & Sustainable Energy Reviews 113, 1-23. https://doi.org/10.1016/j.rser.2019.06.040.

Shu, X., Shen, S.Q., Shen, J.W., Zhang, Y.J., Li, G., Chen, Z., Liu, Y.G., 2021. State of health prediction of lithium-ion batteries based on machine learning: Advances and perspectives. Iscience 24 (11), 1-31. https://doi.org/10.1016/j.isci.2021.103265.

Soltani, S., Kordestani, M., Aghaee, P.K., 2016. New estimation methodologies for well logging problems via a combination of fuzzy Kalman filter and different smoothers. Journal of Petroleum Science and Engineering 145, 704-710. https://doi.org/10.1016/j.petrol.2016.06.032.

Somasundaran, N., Saniee, N.F., Dinh, T.Q., Marco, J., 2023. Study on the Extensibility of Voltage-Plateau-Based Lithium Plating Detection for Electric Vehicles. Energies 16 (6). https://doi.org/10.3390/en16062537.

Song, Y., Sheng, L., Wang, L., Xu, H., He, X., 2021. From separator to membrane: Separators can function more in lithium ion batteries. Electrochemistry Communications 124. https://doi.org/10.1016/j.elecom.2021.106948.

Takyi-Aninakwa, P., Wang, S.L., Zhang, H.Y., Appiah, E., Bobobee, E.D., Fernandez, C., 2022a. A strong tracking adaptive fading-extended Kalman filter for the state of charge estimation of lithium-ion batteries. International Journal of Energy Research 46 (12), 16427-16444.

https://doi.org/10.1002/er.8307.

Takyi-Aninakwa, P., Wang, S.L., Zhang, H.Y., Li, H., Xu, W.H., Fernandez, C., 2022b. An optimized relevant long short-term memory-squared gain extended Kalman filter for the state of charge estimation of lithium-ion batteries. Energy 260. https://doi.org/10.1016/j.energy.2022.125093.

Takyi-Aninakwa, P., Wang, S.L., Zhang, H.Y., Yang, X.Y., Fernandez, C., 2022c. An optimized long short-term memory-weighted fading extended Kalman filtering model with wide temperature adaptation for the state of charge estimation of lithium-ion batteries. Applied Energy 326, 1-21. https://doi.org/10.1016/j.apenergy.2022.120043.

Tan, Q.B., Wang, Z.N., Fan, W., Li, X.D., Li, X.G., Li, F.Q., Zhao, Z.H., 2023. Development Path and Model Design of a New Energy Vehicle in China. Energies 16 (1). https://doi.org/10.3390/en16010220.

Tan, X.J., Zhan, D., Lyu, P.X., Rao, J., Fan, Y.Q., 2021. Online state-of-health estimation of lithium-ion battery based on dynamic parameter identification at multi timescale and support vector regression. Journal of Power Sources 484, 1-10. https://doi.org/10.1016/j.jpowsour.2020.229233.

Tian, H.X., Qin, P.L., Li, K., Zhao, Z., 2020. A review of the state of health for lithium -ion batteries: Research status and suggestions. Journal of Cleaner Production 261, 1-30. https://doi.org/10.1016/j.jclepro.2020.120813.

Tian, J.P., Xiong, R., Lu, J.H., Chen, C., Shen, W.X., 2022. Battery state-of-charge estimation amid dynamic usage with physics-informed deep learning. Energy Storage Materials 50, 718-729. https://doi.org/10.1016/j.ensm.2022.06.007.

Tran, M.-K., Mathew, M., Janhunen, S., Panchal, S., Raahemifar, K., Fraser, R., Fowler, M., 2021. A comprehensive equivalent circuit model for lithium-ion batteries, incorporating the effects of state of health, state of charge, and temperature on model parameters. Journal of Energy

Storage 43. https://doi.org/10.1016/j.est.2021.103252.

Urrea, C., Agramonte, R., 2021. Kalman Filter: Historical Overview and Review of Its Use in Robotics 60 Years after Its Creation. Journal of Sensors 2021. https://doi.org/10.1155/2021/9674015.

Wang, C., Wang, S.L., Zhou, J.Z., Qiao, J.L., 2022a. A Novel BCRLS-BP-EKF Method for the State of Charge Estimation of Lithium-ion Batteries. International Journal of Electrochemical Science 17 (4), 1-17. https://doi.org/10.20964/2022.04.53.

Wang, C., Wang, S.L., Zhou, J.Z., Qiao, J.L., Yang, X., Xie, Y.X., 2023a. A novel back propagation neural network-dual extended Kalman filter method for state-of-charge and state-of-health co-estimation of lithium-ion batteries based on limited memory least square algorithm. Journal of Energy Storage 59, 1-13. https://doi.org/10.1016/j.est.2022.106563.

Wang, C., Xu, M.Y., Zhang, Q.J., Feng, J.H., Jiang, R.Z., Wei, Y., Liu, Y.C., 2021a. Parameters identification of Thevenin model for lithium-ion batteries using self-adaptive Particle Swarm Optimization Differential Evolution algorithm to estimate state of charge. Journal of Energy Storage 44, 1-18. https://doi.org/10.1016/j.est.2021.103244.

Wang, J., Wang, S.L., Yu, C.M., Li, B.W., He, M.F., 2022b. An Online Method for Power State Estimation of Lithium-Ion Batteries under the Constraints of the Fusion Model Considering Temperature Effect. International Journal of Electrochemical Science 17 (7). https://doi.org/10.20964/2022.07.68.

Wang, N., Zhao, Y., Song, T., Zou, X., Wang, E., Du, S., 2022c. Accounting for China's Net Carbon Emissions and Research on the Realization Path of Carbon Neutralization Based on Ecosystem Carbon Sinks. Sustainability 14 (22). https://doi.org/10.3390/su142214750.

Wang, S.L., Cao, J., Xie, Y.X., Gao, H.Y., Fernandez, C., 2022d. A Novel 2-RC Equivalent Model Based on the Self-Discharge Effect for Accurate State-Of-Charge Estimation Of Lithium-Ion Batteries.

International Journal of Electrochemical Science 17 (7). https://doi.org/10.20964/2022.07.60.

Wang, S.L., Fan, Y.C., Jin, S.Y., Takyi-Aninakwa, P., Fernandez, C., 2023b. Improved anti-noise adaptive long short-term memory neural network modeling for the robust remaining useful life prediction of lithium-ion batteries. Reliability Engineering & System Safety 230. https://doi.org/10.1016/j.ress.2022.108920.

Wang, S.L., Fan, Y.C., Yu, C.M., Jin, S.Y., Fernandez, C., Stroe, D.I., 2022e. Improved covariance matching-electrical equivalent modeling for accurate internal state characterization of packing lithium-ion batteries. International Journal of Energy Research 46 (3), 3602-3620. https://doi.org/10.1002/er.7408.

Wang, S.L., Fan, Y.C., Yu, C.M., Jin, S.Y., Takyi-Aninakwa, P., Fernandez, C., Stroe, D.I., 2022f. A novel collaborative multiscale weighting factor-adaptive Kalman filtering method for the time-varying whole-life-cycle state of charge estimation of lithium-ion batteries. International Journal of Energy Research 46 (6), 7704-7721. https://doi.org/10.1002/er.7672.

Wang, S.L., Fernandez, C., Cao, W., Zou, C.Y., Yu, C.M., Li, X.X., 2019a. An adaptive working state iterative calculation method of the power battery by using the improved Kalman filtering algorithm and considering the relaxation effect. Journal of Power Sources 428, 67-75. https://doi.org/10.1016/j.jpowsour.2019.04.089.

Wang, S.L., Fernandez, C., Fan, Y.C., Feng, J.Q., Yu, C.M., Huang, K.F., Xie, W., 2020a. A novel safety assurance method based on the compound equivalent modeling and iterate reduce particle-adaptive Kalman filtering for the unmanned aerial vehicle lithium ion batteries. Energy Science & Engineering 8 (5), 1484-1500. https://doi.org/10.1002/ese3.606.

Wang, S.L., Fernandez, C., Xie, Z.W., Li, X.X., Lou, C.Y., Li, Q., 2019b. A novel weight coefficient calculation method for the real-time state monitoring of the lithium-ion battery packs under the complex current variation working conditions. Energy Science & Engineering 7 (6), 3038-3057. https://doi.org/10.1002/ese3.478.

Wang, S.L., Fernandez, C., Yu, C.M., Fan, Y.C., Cao, W., Stroe, D.I., 2020b. A novel charged state prediction method of the lithium ion battery packs based on the composite equivalent modeling and improved splice Kalman filtering algorithm. Journal of Power Sources 471. https://doi.org/10.1016/j.jpowsour.2020.228450.

Wang, S.L., Fernandez, C., Zou, C.Y., Yu, C.M., Chen, L., Zhang, L., 2019c. A comprehensive working state monitoring method for power battery packs considering state of balance and aging correction. Energy 171, 444-455. https://doi.org/10.1016/j.energy.2019.01.020.

Wang, S.L., Gao, H.Y., Qiao, J.L., Cao, J., Fernandez, C., 2022g. A Novel Square-Root Adaptive Unscented Kalman Filtering Method for Accurate State-of-Charge Estimation of Lithium-ion Batteries. International Journal of Electrochemical Science 17 (7). https://doi.org/10.20964/2022.07.46.

Wang, S.L., Jin, S.Y., Bai, D.K., Fan, Y.C., Shi, H.T., Fernandez, C., 2021b. A critical review of improved deep learning methods for the remaining useful life prediction of lithium-ion batteries. Energy Reports 7, 5562-5574. https://doi.org/10.1016/j.egyr.2021.08.182.

Wang, S.L., Ren, P., Takyi-Aninakwa, P., Jin, S.Y., Fernandez, C., 2022h. A Critical Review of Improved Deep Convolutional Neural Network for Multi-Timescale State Prediction of Lithium-Ion Batteries. Energies 15 (14), 1-27. https://doi.org/10.3390/en15145053.

Wang, S.L., Stroe, D.I., Fernandez, C., Xiong, L.Y., Fan, Y.C., Cao, W., 2020c. A novel power state evaluation method for the lithium battery packs based on the improved external measurable parameter coupling model. Journal of Cleaner Production 242. https://doi.org/10.1016/j.jclepro.2019.118506.

Wang, S.L., Stroe, D.I., Fernandez, C., Yu, C.M., Zou, C.Y., Li, X.X., 2020d. A novel energy management strategy for the ternary lithium batteries based on the dynamic equivalent circuit modeling and differential Kalman filtering under time-varying conditions. Journal of Power Sources 450. https://doi.org/10.1016/j.jpowsour.2019.227652.

Wang, S.L., Takyi-Aninakwa, P., Fan, Y.C., Yu, C.M., Jin, S.Y., Fernandez, C., Stroe, D.I., 2022i. A novel feedback correction-adaptive Kalman filtering method for the whole-life-cycle state of charge and closed-circuit voltage prediction of lithium-ion batteries based on the second-order electrical equivalent circuit model. International Journal of Electrical Power & Energy Systems 139. https://doi.org/10.1016/j.ijepes.2022.108020.

Wang, S.L., Takyi-Aninakwa, P., Jin, S.Y., Yu, C.M., Fernandez, C., Stroe, D.I., 2022j. An improved feedforward-long short-term memory modeling method for the whole-life-cycle state of charge prediction of lithium-ion batteries considering current-voltage-temperature variation. Energy 254. https://doi.org/10.1016/j.energy.2022.124224.

Wang, S.L., Takyi-Aninakwa, P., Yu, C.M., Jin, S.Y., Fernandez, C., 2022k. Improved Compound Correction-Electrical Equivalent Circuit Modeling and Double Transform-Unscented Kalman Filtering for the High-Accuracy Closed-Circuit voltage and State-of-Charge Co-Estimation of Whole-Life-Cycle Lithium-Ion batteries. Energy Technology 10 (12). https://doi.org/10.1002/ente.202200921.

Wang, S.L., Xiong, X., Zou, C.Y., Chen, L., Jiang, C., Xie, Y.X., Stroe, D.I., 2021c. An improved coulomb counting method based on dual open-circuit voltage and real-time evaluation of battery dischargeable capacity considering temperature and battery aging. International Journal of Energy Research 45 (12), 17609-17621. https://doi.org/10.1002/er.7042.

Wang, Y.J., Tian, J.Q., Sun, Z.D., Wang, L., Xu, R.L., Li, M.C., Chen, Z.H., 2020e. A comprehensive review of battery modeling and state estimation approaches for advanced battery management systems. Renewable & Sustainable Energy Reviews 131, 1-18. https://doi.org/10.1016/j.rser.2020.110015.

Wang, Y.T., Wang, S.L., Fan, Y.C., Xie, Y.X., Fernandez, C., 2022l. A Novel Adaptive Back Propagation Neural Network-Unscented Kalman Filtering Algorithm for Accurate Lithium-Ion Battery State of Chargc Estimation. Metals 12 (8), 1-18. https://doi.org/10.3390/met12081369.

Wang, Z., Zhao, X., Zhang, H., Zhen, D., Gu, F., Ball, A., 2023c. Active acoustic emission sensing for fast co-estimation of state of charge and state of health of the lithium-ion battery. Journal of Energy Storage 64. https://doi.org/10.1016/j.est.2023.107192.

Wei, Z., Xiang, B.L., Ting, R.X., 2014. Online parameter identification of the asymmetrical Bouc-Wen model for piezoelectric actuators. Precision Engineering-Journal of the International Societies for Precision Engineering and Nanotechnology 38 (4), 921-927. https://doi.org/10.1016/j.precisioneng.2014.06.002.

Wu, T.Z., Liu, S.Z., Wang, Z.K., Huang, Y.H., 2022. SOC and SOH Joint Estimation of Lithium-Ion Battery Based on Improved Particle Filter Algorithm. Journal of Electrical Engineering & Technology 17 (1), 307-317. https://doi.org/10.1007/s42835-021-00861-y.

Xia, L.L., Wang, S.L., Yu, C.M., Fan, Y.C., Li, B.W., Xie, Y.X., 2022. Joint estimation of the state-of-energy and state-of-charge of lithium-ion batteries under a wide temperature range based on the fusion modeling and online parameter prediction. Journal of Energy Storage 52. https://doi.org/10.1016/j.est.2022.105010.

Xiao, W.J., Wang, S.L., Yu, C.M., Yang, X., Qiu, J.S., Fernandez, C., 2022. Online Parameter Identification and State of Charge Estimation of Lithium-Ion Batteries Based on Improved Artificial Fish Swarms

Forgetting Factor Least Squares and Differential Evolution Extended Kalman Filter. Journal of the Electrochemical Society 169 (12). https://doi.org/10.1149/1945-7111/acaa5b.

Xie, Y.X., Wang, S.L., Fernandez, C., Yu, C.M., Fan, Y.C., Cao, W., Chen, X.P., 2021. Improved gray wolf particle filtering and high-fidelity second-order autoregressive equivalent modeling for intelligent state of charge prediction of lithium-ion batteries. International Journal of Energy Research 45 (13), 19203-19214. https://doi.org/10.1002/er.7014.

Xiong, R., Wang, S., Yu, C., Fernandez, C., Xiao, W., Jia, J., 2023. A novel nonlinear decreasing step-bacterial foraging optimization algorithm and simulated annealing-back propagation model for long-term battery state of health estimation. Journal of Energy Storage 59. https://doi.org/10.1016/j.est.2022.106484.

Xiong, R., Wang, S.L., Fernandez, C., Yu, C.M., Fan, Y.C., Cao, W., Jiang, C., 2021. A Novel Joint Estimation Method of State of Charge and State of Health Based on the Strong Tracking-Dual Adaptive Extended Kalman Filter Algorithm for the Electric Vehicle Lithium-Ion Batteries. International Journal of Electrochemical Science 16 (11). https://doi.org/10.20964/2021.11.18.

Xiong, X., Wang, S.L., Fernandez, C., Yu, C.M., Zou, C.Y., Jiang, C., 2020. A novel practical state of charge estimation method: an adaptive improved ampere-hour method based on composite correction factor. International Journal of Energy Research 44 (14), 11385-11404. https://doi.org/10.1002/er.5758.

Xu, H., Wang, S.L., Fan, Y.C., Qiao, J.L., Xu, W.H., 2022. A novel Drosophila-back propagation method for the lithium-ion battery state of charge estimation adaptive to complex working conditions. International Journal of Energy Research 46 (11), 15864-15880. https://doi.org/10.1002/er.8286.

Xu, J., Cai, X., Cai, S., Shao, Y., Hu, C., Lu, S., Ding, S., 2023. High-Energy Lithium-Ion Batteries: Recent Progress and a Promising Future in Applications. Energy & Environmental Materials. https://doi.org/10.1002/eem2.12450.

Xu, P.P., Li, J.Q., Sun, C., Yang, G.D., Sun, F.C., 2021a. Adaptive State-of-Charge Estimation for Lithium-Ion Batteries by Considering Capacity Degradation. Electronics 10 (2), 1-17. https://doi.org/10.3390/electronics10020122.

Xu, T.T., Peng, Z., Wu, L.F., 2021b. A novel data-driven method for predicting the circulating capacity of lithium-ion battery under random variable current. Energy 218. https://doi.org/10.1016/j.energy.2020.119530.

Xu, W.H., Wang, S.L., Fernandez, C., Yu, C.M., Fan, Y.C., Cao, W., 2020. Novel reduced-order modeling method combined with three-particle nonlinear transform unscented Kalman filtering for the battery state-of-charge estimation. Journal of Power Electronics 20 (6), 1541-1549. https://doi.org/10.1007/s43236-020-00146-z.

Xu, W.H., Wang, S.L., Jiang, C., Fernandez, C., Yu, C.M., Fan, Y.C., Cao, W., 2021c. A novel adaptive dual extended Kalman filtering algorithm for the Li-ion battery state of charge and state of health co-estimation. International Journal of Energy Research 45 (10), 14592-14602. https://doi.org/10.1002/er.6719.

Xue, A.R., Yang, W.L., Yuan, X.M., Yu, B.P., Pan, C.F., 2022. Estimating state of health of lithium-ion batteries based on generalized regression neural network and quantum genetic algorithm. Applied Soft Computing 130. https://doi.org/10.1016/j.asoc.2022.109688.

Yang, K., Chen, Z., He, Z., Wang, Y., Zhou, Z., 2020. Online estimation of state of health for the airborne Li-ion battery using adaptive DEKF-based fuzzy inference system. Soft Computing 24 (24), 18661-18670. https://doi.org/10.1007/s00500-020-05101-5.

Yang, S.J., Zhang, C.P., Jiang, J.C., Zhang, W.G., Zhang, L.J., Wang, Y.B., 2021. Review on state-of-health of lithium-ion batteries: Characterizations, estimations and applications. Journal of Cleaner Production 314, 1-21. https://doi.org/10.1016/j.jclepro.2021.128015.

Yang, X., Wang, S.L., Xu, W.H., Qiao, J.L., Yu, C.M., Fernandez, C., 2022a. Fuzzy adaptive singular value decomposition cubature Kalman filtering algorithm for lithium-ion battery state-of-charge estimation. International Journal of Circuit Theory and Applications 50 (2), 614-632. https://doi.org/10.1002/cta.3166.

Yang, X., Wang, S.L., Xu, W.H., Qiao, J.L., Yu, C.M., Takyi-Aninakwa, P., Jin, S.Y., 2022b. A novel fuzzy adaptive cubature Kalman filtering method for the state of charge and state of energy co-estimation of lithium-ion batteries. Electrochimica Acta 415. https://doi.org/10.1016/j.electacta.2022.140241.

Yang, X.Y., Liu, J.X., Pei, N.B., Chen, Z.Q., Li, R.Y., Fu, L.J., Zhang, P., Zhao, J.B., 2023. The Critical Role of Fillers in Composite Polymer Electrolytes for Lithium Battery. Nano-Micro Letters 15 (1). https://doi.org/10.1007/s40820-023-01051-3.

Yao, X.-Y., Chen, G., Pecht, M., Chen, B., 2023. A novel graph-based framework for state of health prediction of lithium-ion battery. Journal of Energy Storage 58. https://doi.org/10.1016/j.est.2022.106437.

Yu, P., Wang, S.L., Yu, C.M., Jiang, C., Shi, W.H., 2022a. An adaptive fractional-order extended Kalman filtering for state of charge estimation of high-capacity lithium-ion battery. International Journal of Energy Research 46 (4), 4869-4878. https://doi.org/10.1002/er.7480.

Yu, P., Wang, S.L., Yu, C.M., Shi, W.H., Li, B.W., 2022b. Study of hysteresis voltage state dependence in lithium-ion battery and a novel asymmetric hysteresis modeling. Journal of Energy Storage 51. https://doi.org/10.1016/j.est.2022.104492.

Yu, Z., Liu, N., Zhang, Y., Qi, L., Li, R., 2023. Battery SOH Prediction Based on Multi-Dimensional Health Indicators. Batteries-Basel 9 (2). https://doi.org/10.3390/batteries9020080.

Yuan, M., Liu, K., 2020. Rational design on separators and liquid electrolytes for safer lithium-ion batteries. Journal of Energy Chemistry 43, 58-70. https://doi.org/10.1016/j.jechem.2019.08.008.

Zahoor, A., Yu, Y.J., Zhang, H.L., Nihed, B., Afrane, S., Peng, S., Sapi, A., Lin, C.J., Mao, G.Z., 2023. Can the new energy vehicles (NEVs) and power battery industry help China to meet the carbon neutrality goal before 2060? Journal of Environmental Management 336. https://doi.org/10.1016/j.jenvman.2023.117663.

Zhang, C., Amietszajew, T., Li, S., Marinescu, M., Offer, G., Wang, C., Guo, Y., Bhagat, R., 2022a. Real-time estimation of negative electrode potential and state of charge of lithium-ion battery based on a half-cell-level equivalent circuit model. Journal of Energy Storage 51, 1-10. https://doi.org/10.1016/j.est.2022.104362.

Zhang, C.Y., Wang, S.L., Yu, C.M., Xie, Y.X., Fernandez, C., 2022b. Improved Particle Swarm Optimization-Extreme Learning Machine Modeling Strategies for the Accurate Lithium-ion Battery State of Health Estimation and High-adaptability Remaining Useful Life Prediction. Journal of the Electrochemical Society 169 (8). https://doi.org/10.1149/1945-7111/ac8a1a.

Zhang, C.Y., Wang, S.L., Yu, C.M., Xie, Y.X., Fernandez, C., 2022c. Novel Improved Particle Swarm Optimization-Extreme Learning Machine Algorithm for State of Charge Estimation of Lithium-Ion Batteries. Industrial & Engineering Chemistry Research 61 (46), 17209-17217. https://doi.org/10.1021/acs.iecr.2c02476.

Zhang, F., Xing, Z.X., Wu, M.H., 2023a. State of health estimation for Li-ion battery using characteristic voltage intervals and genetic algorithm optimized back propagation neural network. Journal of Energy Storage 57. https://doi.org/10.1016/j.est.2022.106277.

Zhang, L., Li, X., Yang, M., Chen, W., 2021a. High-safety separators for lithium-ion batteries and sodium-ion batteries: advances and perspective. Energy Storage Materials 41, 522-545. https://doi.org/10.1016/j.ensm.2021.06.033.

Zhang, L., Wang, S.L., Stroe, D.I., Zou, C.Y., Fernandez, C., Yu, C.M., 2020. An Accurate Time Constant Parameter Determination Method for the Varying Condition Equivalent Circuit Model of Lithium Batteries. Energies 13 (8). https://doi.org/10.3390/en13082057.

Zhang, L., Wang, S.L., Zou, C.Y., Fan, Y.C., Jin, S.Y., Fernandez, C., 2021b. A novel streamlined particle-unscented Kalman filtering method for the available energy prediction of lithium-ion batteries considering the time-varying temperature-current influence. International Journal of Energy Research 45 (12), 17858-17877. https://doi.org/10.1002/er.6930.

Zhang, M.Y., Wang, S.L., Yang, X., Xie, Y.X., Liu, K., Zhang, C.Y., 2023b. Improved Backward Smoothing-Square Root Cubature Kalman Filtering and Variable Forgetting Factor-Recursive Least Square Modeling Methods for the High-Precision State of Charge Estimation of Lithium-Ion Batteries. Journal of the Electrochemical Society 170 (3). https://doi.org/10.1149/1945-7111/acb10b.

Zhang, M.Y., Wang, S.L., Yang, X., Xu, W.H., Yang, X.Y., Fernandez, C., 2022d. A Novel Square Root Adaptive Unscented Kalman Filter Combined with Variable Forgetting Factor Recursive Least Square Method for Accurate State-of-charge Estimation of Lithium-Ion Batteries. International Journal of Electrochemical Science 17 (9). https://doi.org/10.20964/2022.09.27.

Zhang, Q., Wang, D., Schaltz, E., Stroe, D.-I., Gismero, A., Yang, B., 2023c. Lithium-ion battery calendar aging mechanism analysis and impedance-based State-of-Health estimation method. Journal of Energy Storage 64. https://doi.org/10.1016/j.est.2023.107029.

Zhang, X., Hou, J., Wang, Z., Jiang, Y., 2023d. Joint SOH-SOC Estimation Model for Lithium-Ion Batteries Based on GWO-BP Neural Network. Energies 16 (1). https://doi.org/10.3390/en16010132.

Zhang, X., Sun, J., Shang, Y., Ren, S., Liu, Y., Wang, D., 2022e. A novel state-of-health prediction method based on long short-term memory network with attention mechanism for lithium-ion battery. Frontiers in Energy Research 10. https://doi.org/10.3389/fenrg.2022.972486.

Zhang, Y., Nguyen, R.T., Liaw, B., 2021c. Status and Gap in Rechargeable Lithium Battery Supply Chain: Importance of Quantitative Failure Analysis. Proceedings of the Ieee 109 (6), 1029-1038. https://doi.org/10.1109/jproc.2020.3047880.

Zhao, J.H., Zhu, Y., Zhang, B., Liu, M.Y., Wang, J.X., Liu, C.H., Hao, X.W., 2023. Review of State Estimation and Remaining Useful Life Prediction Methods for Lithium-Ion Batteries. Sustainability 15 (6), 1-22. https://doi.org/10.3390/su15065014.

Zhao, X.B., Qian, X., Xuan, D.J., Jung, S.H., 2022. State of charge estimation of lithium-ion battery based on multi-input extreme learning machine using online model parameter identification. Journal of Energy Storage 56, 1-13. https://doi.org/10.1016/j.est.2022.105796.

Zheng, H.H., Li, X.X., Huang, Z., Yang, X.Q., Deng, J.H., Zhang, G.Q., Tang, X.W., 2022a. Investigation of the Performance and Recession Mechanisms of High-Nickel Ternary Lithium-Ion Batteries Under Artificial Aging Discharge Rates. Energy Technology 10 (11). https://doi.org/10.1002/ente.202200600.

Zheng, L.C., Chen, G., Liu, L.T., Hu, Y.Q., 2022b. Tracing of lithium supply and demand bottleneck in China's new energy vehicle industry-based on the chart of lithium flow. Frontiers in Energy Research 10. https://doi.org/10.3389/fenrg.2022.992617.

Zhou, H., Wang, S.L., Yu, C.M., Xia, L.L., Fernandez, C., 2022. Research on SOC Estimation for Lithium ion batteries Based on Improved PNGV Equivalence Model and AF-UKF Algorithm. International

Journal of Electrochemical Science 17 (8). https://doi.org/10.20964/2022.08.31.

Zhou, W.L., Zheng, Y.P., Pan, Z.J., Lu, Q., 2021. Review on the Battery Model and SOC Estimation Method. Processes 9 (9), 1-23. https://doi.org/10.3390/pr9091685.

Zhu, F., Fu, J.Q., 2021. A Novel State-of-Health Estimation for Lithium-Ion Battery via Unscented Kalman Filter and Improved Unscented Particle Filter. Ieee Sensors Journal 21 (22), 25449-25456. https://doi.org/10.1109/jsen.2021.3102990.

Zou, C.N., Ma, F., Pan, S.Q., Lin, M.J., Zhang, G.S., Xiong, B., Wang, Y., Liang, Y.B., Yang, Z., 2022. Earth energy evolution, human development and carbon neutral strategy. Petroleum Exploration and Development 49 (2), 468-488. https://doi.org/10.1016/s1876-3804(22)60040-5.